The American Electro Magnetic Telegraph

With the Report[...] of All
Telegraphs Know[...]vanism

Alfred Vail

Alpha Editions

This edition published in 2024

ISBN : 9789366388069

Design and Setting By
Alpha Editions
www.alphaedis.com
Email - info@alphaedis.com

INTRODUCTION.

The propriety of presenting to the public a work of this character, seemed desirable, from the frequent calls made upon the author for some accurate and full description of the American Electro Magnetic Telegraph, which might assist to an intelligible comprehension of the principles upon which it is based, and the mode of its operations, as well as descriptions of those plans now in operation in Europe. In the execution of this task it has been his determination to spare no labour, and to omit nothing that could enable those who had never seen the operation of the telegraph, to obtain a full understanding, of the subject, and also to judge for themselves of the merit of the American invention, as compared with those of Europe. For this purpose eighty-one wood cuts are introduced to illustrate this and collateral subjects.

The various reports of Congress which have been made, from time to time, as the subject of the Electro Magnetic Telegraph has been presented to them, have been embraced in the work. They contain much information in relation to the origin and progress of the invention, as well as other useful matter. In the closing part of the work is given a synopsis of the early discoveries in electricity; the experiment of Franklin, and also the discoveries of ingenious and scientific gentlemen of the present day. The principal part, however, is devoted to a full and complete description of the various plans of telegraphic communication, by means of electricity and galvanism, in [Pg viii]the chronological order of their invention; by which it will be seen, that for priority as well as originality, America has the pre-eminence, not only at the time of the invention, but up to the present period; nothing having yet been brought forward that fulfils so completely the conditions of what is signified by the term *telegraph*, as that plan invented by Professor Morse. Some of the foreign plans the author has found extremely difficult to illustrate, without almost re-inventing them, so imperfectly and obscurely have they been described.

The experimental line from Washington to Baltimore has been in successful operation for more than a year, and has been the means of conveying much important information: consisting of messages to and from merchants, members of Congress, officers of the government, banks, brokers, police officers; parties, who by agreement had met each other at the two stations, or had been sent for by one of the parties; items of news, election returns, announcement of deaths, inquiries respecting the health of families and individuals, the daily proceedings of the Senate and House of Representatives, orders for goods, inquiries respecting the sailing of vessels, proceedings of cases in the various courts, summoning of witnesses, messages in relation to special and express trains, invitations, the receipt of money at one station and its payment at the other, for persons requesting the transmission of funds from debtors, consultation of physicians, and messages of every character usually sent by mail.

The author trusts that the work will be received as one of a practical character, and furnish to those desirous to acquaint themselves with the subject, such information as they seek.

ALFRED VAIL.

Washington, D. C. *August 18, 1845.*

THE ELECTRO MAGNETIC TELEGRAPH.

THE GALVANIC BATTERY.

The galvanic battery, the generator of that subtle fluid, which performs so important a part in the operation of the Electro Magnetic Telegraph, is as various in its form and arrangement, as the variety of purposes to which it is applied. They all, however, involve the same principle. It is not our design here to describe the various modes of constructing it, but to confine our remarks more immediately to that used for the Telegraph.

The effects produced by the galvanic fluid upon the metallic bodies, iron and steel, exciting in them the power of attraction or magnetism, its decomposing effects upon liquids, resolving them into their simple elements, its effects upon the animal system, in producing a spasmodic and sudden irritation, are generally well known. But of the character of the fluid itself, its own essence or substance, we know nothing. In some of its phenomena, it resembles the electricity of the heavens; both find a conductor in the metals; both exhibit a spark, and both are capable of producing shocks, or when applied, cause the animal system to be sensible to them. Again, in other of its phenomena it is totally unlike it. The galvanic fluid is essentially necessary in producing the electro magnet; while the electricity of the heavens, or as it is generally termed, *machine electricity*, has no such power for practical purposes. The former is more dense, so to speak, and more easily confined to its conductors, while the latter becomes dissipated and lost in the atmosphere long before it has reached the opposite extremity of a long conductor. The former is continuous in its supply; while the latter is at irregular intervals. The former always needs a continuous conductor; while the latter will pass from one metallic conductor to another without that connection. The latter would not subserve the purposes required in the working of the Electro Magnetic Telegraph, and as it is neither essential nor antagonistical, its presence upon the galvanic conductors or wires, at the same time those wires are being used for telegraphic communication, does in no way interrupt or confuse its operation; and its presence is only known from the suddenness of its discharge at long intervals, accompanied by a bright spark, with a loud crack, like that of a coachman's whip.

The most simple mode of developing the galvanic fluid is in the following manner: if a common glass tumbler is two-thirds filled with dilute muriatic acid, and a piece of bright zinc, five inches long and one inch wide, immersed in the liquid, at one of its ends, slight action will be discovered upon it. If a slip of copper be then taken, of the same dimensions, and one end immersed in the liquid, but separated from that portion of the zinc immersed, and not permitted to touch it; and the two projecting ends of the zinc and copper, above the liquid be brought in contact, an active decomposition of the muriatic acid will appear.

While the two outer ends are in contact, there is that current formed in the metallic plates, which is termed galvanic. If the contact is broken, the action ceases; if it is again renewed, the action is recommenced. Another very simple experiment, and within the power of every one to demonstrate for themselves, is that of applying a piece of zinc to the underside of the tongue, and to the upperside, a silver coin, and then by bringing their projecting ends in contact, a sensible and curious effect is experienced upon the tongue. It is a feeble galvanic shock, and is proof of the presence of that fluid termed galvanic.

We will now proceed to describe the battery used for telegraphic purposes; the same in principle, but in arrangement more complicated, and far more powerful than those in common use. Two distinct acids are employed; two metals and two vessels. Each part will be described separately, and then the whole, as put together ready for use.

First. A glass tumbler of the ordinary size is used, or about three inches high and two inches and three quarters in diameter.

Second. The zinc cylinder, made of the purest zinc, and cast in an iron mould, represented by figure 1.

Fig. 1.

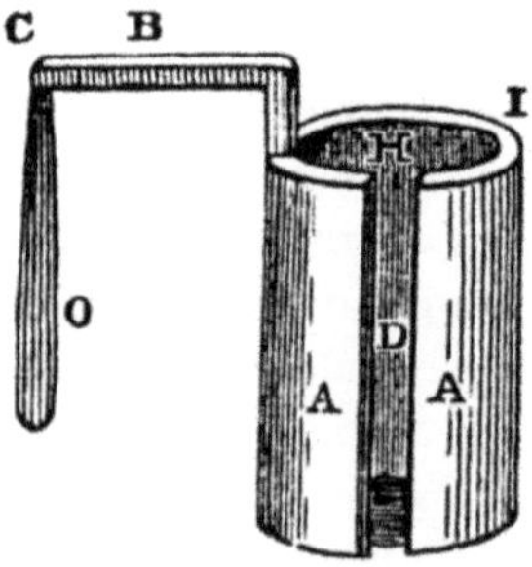

It is three inches high, and two inches in diameter. The shell I is three-eighths of an inch in thickness. D is an opening in the cylinder, parallel with its axis, and is of no other use than to aid in the operation of casting them, and facilitating the access of the fluid to the interior. A A represents the body of the cylinder. B is a projecting arm, first rising vertically from the shell, and then projecting horizontally one and three quarters of an inch. To this arm, at C, is soldered a platinum plate of the thickness of tin foil, and hanging vertically from the arm B, as seen at O, and of the form shown in the figure.

This constitutes the zinc cylinder and platinum plate, the two metals used in the battery.

Fig. 2.

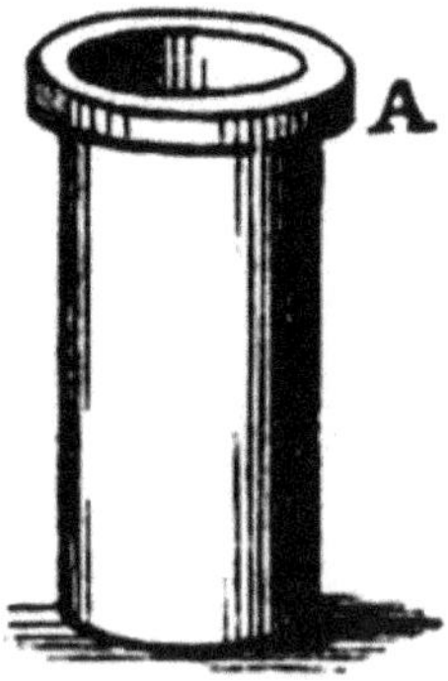

Third. The porous cup. To avoid an erroneous impression in the use of the term porous, it will suffice to state, that it is a cup of the form represented by figure 2, made of the same materials as stone-ware, and baked without being *glazed.*[1] A represents the rim surrounding the top. From the under side of the rim to the bottom, it is three inches long, and one and one-quarter in diameter. The rim projects one-quarter of an inch, and the shell of the cup is one-eighth of an inch thick. These several parts are placed together thus. The porous cup, fig. 2, is set in the hollow of the zinc cylinder, fig. 1, represented by H, with the rim of the cup resting upon the top of the zinc at I. The zinc cylinder is then placed in the glass tumbler. The whole is represented in figure 3.

Fig. 3.

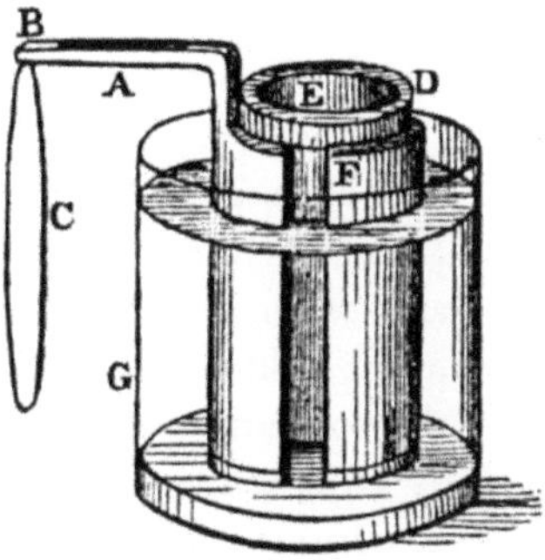

D represents the porous cup, F the zinc cylinder, G the glass tumbler, A the projecting arm of the zinc, C the platinum plate, and B the over-lapping of the platinum plate upon the zinc arm, where it is soldered to it. It is now in a condition to receive the acids, which are two: first, pure nitric acid, and second, sulphuric acid, diluted in the proportion of one part of sulphuric acid to twelve of water. First fill the porous cup with the nitric acid, to within one-quarter of an inch of the top; then fill the glass with the diluted sulphuric acid, till it reaches to a level with the nitric acid in the porous cup. One glass of the battery is now ready for use, and as all the other members of the battery are similarly constructed, (there being many or few, as circumstances require,) and are to be prepared and filled with their appropriate acids in the same manner, the above description will suffice. There remains, however, some further explanation in regard to the extremities of the series of glasses, that is, the mode of connecting the zinc of the first glass with the wire leading from it, and also the mode of connecting the platinum of the last glass with the wire leading from that end of the series of glasses. Figure 4 represents their arrangement.

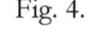

Fig. 4.

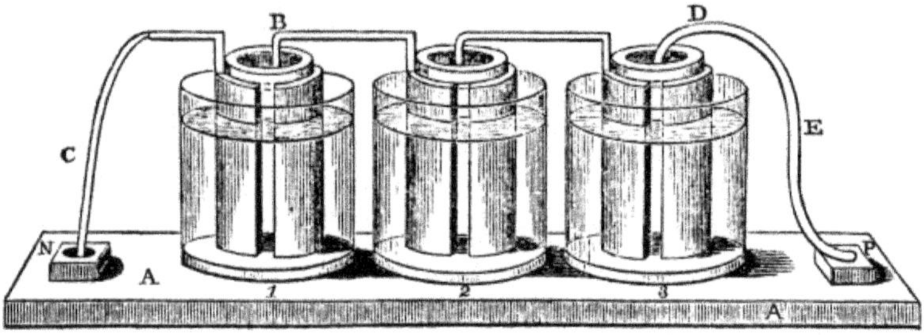

The glasses being all separately supplied with their acids, and otherwise prepared, they are put together upon a table, A A, perfectly dry, and made of hard wood. The first member of the series has soldered to its zinc arm a strip of copper, C, which, extending downward, has its end, previously brightened and amalgamated, immersed in a cup of mercury at N. The cup being permanently secured to the table. Then the second glass is taken, and the platinum, B, at the end of the zinc arm, is gently let fall into the porous cup, so that it shall be in the centre of the cup, and reaching down as far as its length, when the glass rests upon the table. The third glass is then taken and placed in the same manner, and so on to the last. The last glass has, in its porous cup, the platinum plate, D, soldered to a strip of copper, E, which is so constructed as to turn at the top, and admit of the easy introduction of the platinum into the porous cup, while the other end of the copper, previously prepared like the copper of the other end of the battery, terminates in a cup of mercury, P. The cup being capable of adjustment, so as to bring the platinum directly over the porous cup; is, when adjusted, secured permanently to the table. The battery, thus arranged, is ready to be applied.

THE WIRE.

The wire used in making helices for the magnets, and for connecting the telegraphic stations, is made of copper of the best quality, and annealed. It is covered with cotton thread, so as to conceal every part of the metallic surface, not so much to prevent corrosion or waste from the action of the atmosphere, as to prevent a metallic contact of one wire with another, when placed near each other. After the wire is covered, it is then saturated with shellac, and then, again, with a composition of asphaltum, beeswax, resin and linseed oil. It is now in a condition to be extended upon the poles. That portion of the wire of which the helices are made is only saturated with shellac.

THE ELECTRO MAGNET.

The electro magnet is the basis upon which the whole invention rests in its present construction; without it, it would entirely fail. As it is of so much importance, a detailed account will be given of the construction of the electro magnet, as used for telegraphic purposes. A bar of soft iron, of the purest and best quality, is taken and made into the form presented in figure 5, which consists of four parts, viz. A F and A F are the two legs or prongs of the magnet,[2] of a rounded form, and bent at the top, approaching each other towards the centre, where the ends of each prong, without touching, turn up, and present flat, smooth and clean surfaces, level with each other at F F. The other end of these prongs or legs is turned smaller than the body, on the end of which is a screw and nut, C C. These ends pass through a plate of iron, B, of the same quality, at I and I, until they rest upon the plate at the shoulder produced by turning them smaller. They are then both permanently secured to the plate, B, by the nuts, C C, and the whole becomes as one piece. This arrangement is made for the purpose of putting on the coils or taking them off with facility. The form most common for electro magnets is that of the horse-shoe; and is simply a bar of iron bent in that form. E represents a small flat plate of soft iron, sufficiently large to cover the faces of the two prongs, F and F, presenting on its under side a surface clean and smooth, and parallel with the faces, F and F.

Fig. 5.

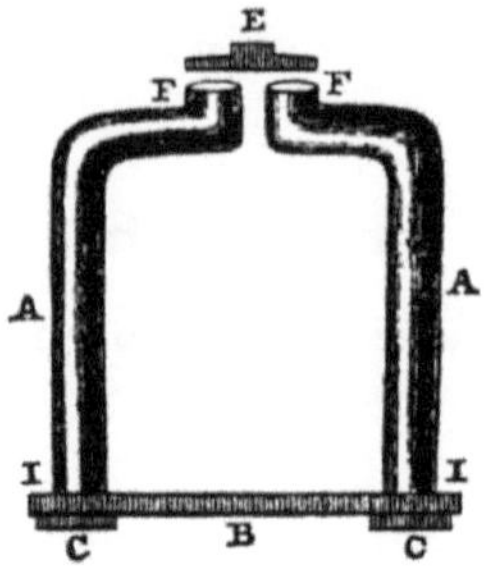

The coils or helices of wire, which surround the prongs, A A, necessary to complete the electro magnet, consist of many turns of wire, first running side by side, covering the form upon which the spiral is made, until the desired length of the coil is obtained; the wire is then turned back, and wound upon the first spiral, covering it, until the other end of the coil is reached, where the winding began; then again mounting upon the second spiral, covers it, and in the same manner it is wound back and forth, until the required size of the coil is attained.

Fig. 6.

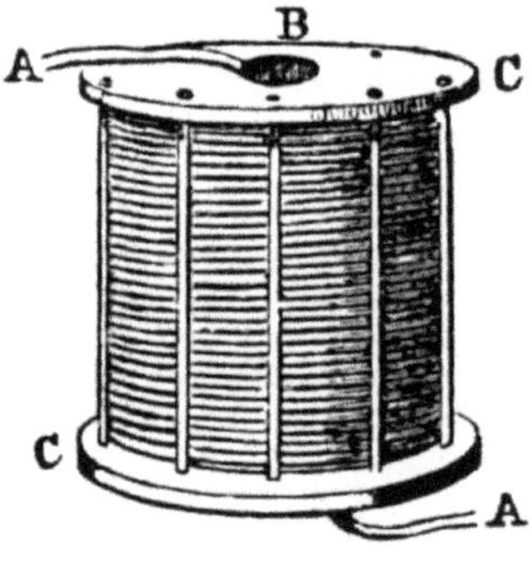

The coil is wound upon a form of the size (or a little larger) of the legs of the magnet, and when the coil is completed, the form is taken out, leaving an opening in the centre, B, into which the prongs may freely pass. Figure 6 represents a coil constructed in the manner described. A and A are the two ends of wire which are brought out from the coils. The one proceeds from the centre of the coil, and the other from the outside. C and C are circular wooden heads, on each end of the coil, and fastened to it by binding wire, running from one head to the other, around the coil. The wire used in constructing it, as heretofore mentioned, is covered in the same manner as bonnet wire, and saturated or varnished with gum shellac. This preparation is necessary, in order to prevent a metallic contact of the wires with each other. Such a contact of some of the wires with others encircling the iron prong, would either weaken or altogether destroy the effect intended by their many turns. If the wires were bare, instead of being covered, the galvanic fluid, when applied to the two ends, A and A, instead of passing through the whole length of the wire in the coil as its conductor, would pass laterally through it as a mass of copper, in the shortest direction it could take. For this reason, they require a careful and most perfect insulation. Two coils are thus prepared for each magnet, one for each prong, A and A, figure 5.

Fig. 7.

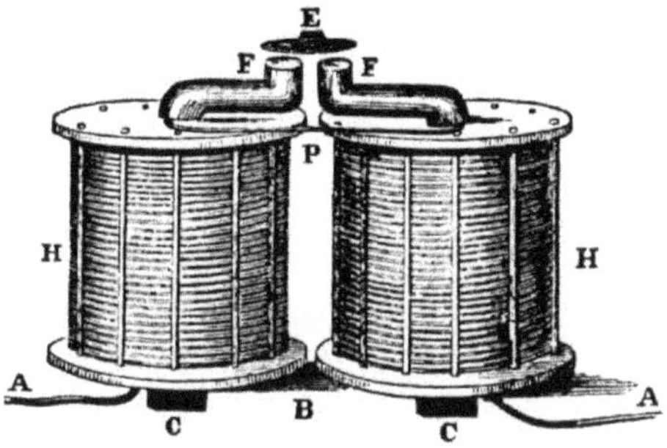

Figure 7 exhibits a view of the magnet; figure 5, with its two coils, H and H, placed upon the prongs. Those parts of the magnet, not concealed by the coils, are lettered as in figure 5, and correspond with its description. P represents the wire connecting the coil H with H, and A and A the ends of the wires leaving the coils.

We now proceed to explain the manner by which the magnet is secured upon a frame, and the arrangement of the armature, E, figure 7, upon a lever, so that the motion peculiar for telegraphic writing may be shown.

Fig. 8.

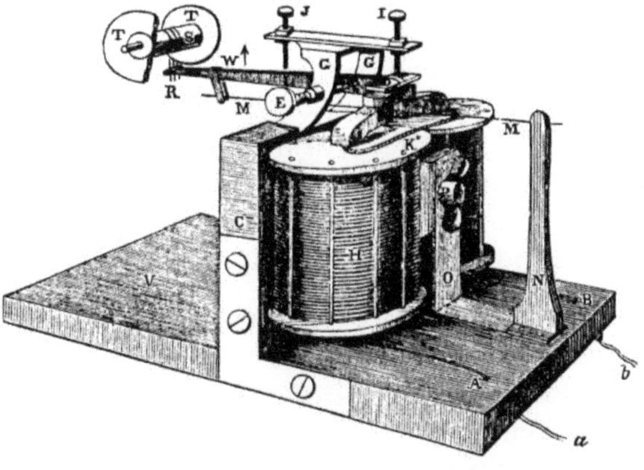

Figure 8 exhibits, in perspective, a view of the electro magnet and the pen lever, in a condition to show the effect of the galvanic battery upon the prongs of the magnet, F and F, and the armature, D, and the movement of the pen lever to which the electro magnet is secured. A bolt, upon the end of which is a head or shoulder, passes through the centre of the upright block, C, and between the coils, H and H, and also through the brass brace, O, projecting a little beyond it, with a screw cut upon its end. The thumb-nut, P, fitted to it, is then put on, and the whole firmly held by screwing the thumb-nut as far as possible. F and F are the faces of the iron prongs, as shown in figure 7, presenting their flat surface to the armature, D. L is the pen lever, suspended upon steel points, as its axis, which pass through its side at X, and soldered to it. Each end of this steel centre is tapered so as to form a sharp and delicate point or pivot. E is a screw, passing through the side of the brass standard, G, and presenting at its end a sunken centre, the reverse of the steel pivot point at X. There is also another screw, similar to E, passing through the other side of the standard at G', with a sunken centre in its end. By the extremities of these two screws, to which the tapered ends of the steel centre is fitted, the pen lever is suspended, so as delicately to move up and down, as shown by the direction of the arrow. The brass standard, G, is secured to the upright block, C. D is the armature, soldered to the end of the brass pen lever, L, separated from the faces of the magnet, F and F, about the eighth of an inch. W is a yoke, secured to the lever by a screw, and which admits through its lower part the steel wire spring, M M, for the purpose of bringing down the lever when not acted upon by the electro magnet. The spring is secured to a brass standard at the top, represented by N. R represents the three steel points of the pen,[3] which mark upon the paper the telegraphic characters; each of which strike into its own appropriate groove in the steel roller, S. T and T are the flanges of the steel roller, S, and which confine the paper as it passes between the pen points, R, and the steel roller, S, described more fully hereafter. J and I are two screws in the horizontal cross bar attached to the standard, G, and are used for the purpose of adjusting and limiting the pen lever in its movement upward and downward; the one to prevent the pen points from striking too deeply into the paper and tearing it, and the other to prevent the armature from receding too far from the faces of the electro magnet, and beyond its attraction, when it is a magnet. K is the connecting wire of the two coils H and H. A and B show the ends of the wire, one coming from each coil and passing through the stand, and seen below at *a* and *b*.

Having explained this arrangement of the electro magnet, the pen lever, and the battery; the effect of the latter upon the former will now be described. Let one of the wires from the coils, figure 8,—*a*, for instance, be extended so far, that it can conveniently and securely be connected with the mercury cup, N, figure 4, of that pole of the battery. Then take the wire *b*, figure 8, and extend it also to a convenient length, so as to be freely handled, and connect it with the mercury cup, P, figure 4, of the other pole of the battery. It will be found at the instant the connection is made, that the lever, L, figure 8, will fly up in the direction of the arrow at W. The iron prongs in the centre of the coils, H and H, which were before perfectly free from any attractive power, have now become powerfully magnetic by the inductive influence of the galvanic current following the circuitous turns of the wire around the iron, so that now the electro magnet is capable of sustaining twenty or twenty-five pounds weight. This magnetic power concentrated in the faces of the electro magnet, F and F, attracts to it the armature or small iron, D, drawing the pen lever down on that side of its axis, and producing a reverse motion on the other side at L. Now take out the wire *b* from the mercury cup, and in an instant its magnetism is gone, and the lever, L, falls by the action of the spring, M. If the circuit is closed a second time, the lever again flies up; and if immediately broken, falls. In this manner it will continue to operate in perfect obedience to the closing or breaking of the circuit. If the circuit is closed and broken in rapid succession, the lever obeys and exhibits a constant and rapid vibration. If the circuit is closed and then broken after a short interval, the lever will remain up the same length of time, the circuit is closed, and falls upon its being broken. Whatever may be the time the circuit is broken, the lever will remain up for the same length of time, and whatever may be the time it continues broken, the lever will remain down for the same time. Suppose the magnet is separated at the distance of one mile from the battery; upon manipulating at the battery, at that distance, in the manner just described, the same vibratory motion is produced in all its varieties, as when they were removed only a short distance. Separate them 10 miles, and still the same mysterious fluid is obedient to the pleasure of the operator in producing the desired motion of the pen lever. If they were separated at distances of 100 or 1000 or 100,000 miles apart, the lever would doubtless obey the manipulations of the operator, as readily as if only distant a few feet. Here is exhibited the principle upon which Morse's Electro Magnetic Telegraph is based, and which gives to the several portions of the civilized world the power of holding instantaneous communication with each other, with a rapidity far beyond what has ever before been attained. As the above explanation is given only in reference to the power of the electro magnet, when connected with the battery, and to show the movements of the pen lever, we shall speak of the arrangement of the wires for extended lines hereafter.

Having now explained the electro magnet and its operation through the agency of the battery, we will proceed to describe those various parts of the register, by which the electro magnet is made subservient to the transmission of intelligence from one distant point to another.

Figure 9 represents, in perspective, the whole of the register, as also the key or correspondent. The electro magnet, H and H, and the pen lever, L, which have just been described under figure 8, need not be recapitulated here. The letters used in figure 8, represent the same parts of the electro magnet in this figure.

The brass frame containing the clock work, or rather wheel work, of the instrument, is seen at 5 and 5. The whole purpose of the clock work is to draw the paper,[4] 2, 2, under the steel roller, S, and over the pen, R, at an uniform rate.

There is also an arrangement in connection with the wheel work, by means of which the clock work is put in motion and stopped at the pleasure of the operator at the distant station. How this is done will now be explained. Upon the shaft, R', is a brass barrel, upon which is wound the cord to which the weight, 4, is suspended, and by means of which and the intermediate wheels, the motion produced, is communicated to two rollers (not seen in this figure, see fig. 10, E F) in advance of the steel grooved roller, S. These two rollers grasp the paper, 2, 2, 3, between them, and supply it to the pen at a given and uniform rate; the rate being determined by the adjustment of the wings of the fly, connected with the train.

Fig. 9.

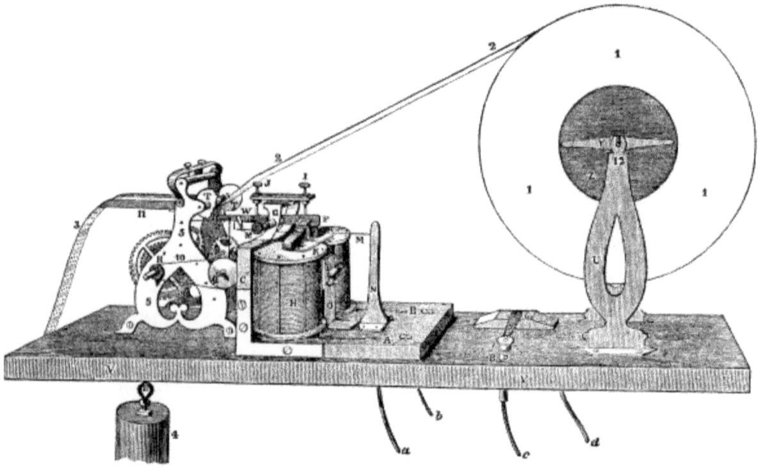

Fig. 10.

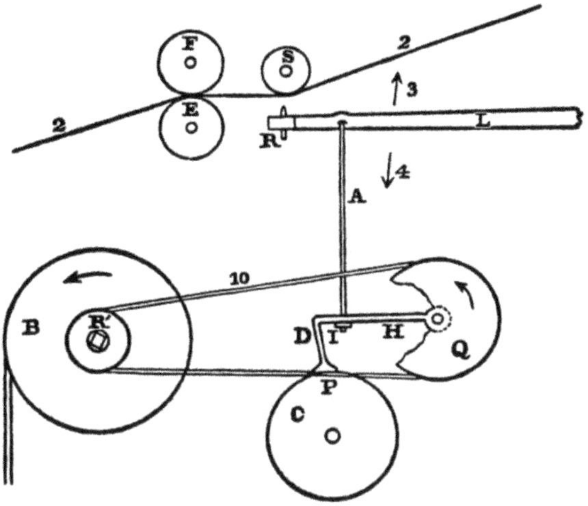

We will now describe, by figure 10, those parts connected with the wheel work, which could not be easily shown in figure 9. F and E represent, in outline, the two rollers which grasp the paper, 2 and 2. The roller E is connected with the train by a cog wheel upon it. F is not so connected; but is pressed hard upon E by means of springs upon the ends of the axle; S represents the grooved steel roller beneath which the paper, 2 and 2, is seen to pass. Directly under the steel roller is one of the steel pen points at R, upon the end of the pen lever; a part of which only is shown. Thus far the description given of the clock work, relates to those parts, by the agency of which the pen is supplied with paper. We now proceed to explain that part connected with the clock and pen lever, by which the clock is set in motion or stopped at the option of the distant operator.

In figure 9, at R', is seen a small pulley upon the barrel shaft of the clock work; at Q, is another pulley, but larger. From the pulley, R', is a cord,[5] or band, 10, proceeding to pulley, Q, and then returning under it to pulley, R', making it continuous. This band communicates the motion of pulley, R', to the pulley, Q. In figure 10, these pulleys are represented by the same letters. B represents the barrel; the arrow, the direction in which it revolves when in motion. The arrow at Q shows the direction which it takes when motion is communicated to it by R'. Part of the pulley, Q, is broken away in order to show the arm, H, soldered at the middle of the same spindle upon which is the pulley, Q, and directly beneath the pen lever, L. It is bent at D, so as to turn down and strike the wooden friction wheel, C, at the point, P. The friction wheel is secured upon the last spindle of the train at its middle and directly under the lever, L. From the pen lever, L, is seen a small rod of wire, A, passing down through the arm, H, with screw and nut under it, at I, for the purpose of shortening or lengthening it. It is permitted to work free, both at its connection with the lever and arm. This wire is also extended and passes down through the platform, where it operates upon a hammer for striking a bell, to apprise the operator that a communication is to be sent. The several parts being now explained, their combined action is as follows:

The arm, H and D, is a break, which when brought in contact with the friction wheel, C, prevents the weight of the clock work from acting upon the train, and there is no motion. By the action of the magnet, the pen lever, L, is carried up in the direction of the arrow, 3, and takes with it the connecting rod, A, and also the break, H, D. The break being thus removed from the friction wheel, C, the clock work commences running by the power of the weight. The barrel, B, must consequently turn in the direction of the arrow upon it; this motion is communicated by the band to Q, which revolves in the direction of its arrow; consequently, if the lever, L, is not still held up by the magnet, the break is descending slowly; and when it reaches P, stops the motion of the clock train, unless the pen lever continues in motion, in which case the break, D, is kept up from the friction wheel, thus permitting the clock work to run, until the lever ceases to move, when the break is gradually brought down upon the friction wheel, and the train stops. By this contrivance, the operator at a distance can so control the movement of the paper at the remote register, that when he wishes to write, it shall be put in motion, his pen be supplied with paper, and when he has finished his writing, the register shall stop.

U represents (figure 9) the brass standards, one on each side of the large roll of paper, 1, 1, 1, which it supports. Z is a wooden hub, upon which the roll is placed; and 12, the steel arbor of the hub, and upon which the whole easily revolves as the paper, 2 and 2, is drawn off by the clock work. Y is a brass spring, between the hub and the standard; and keeps the paper stretched between the roll and the pen.

Fig. 11.

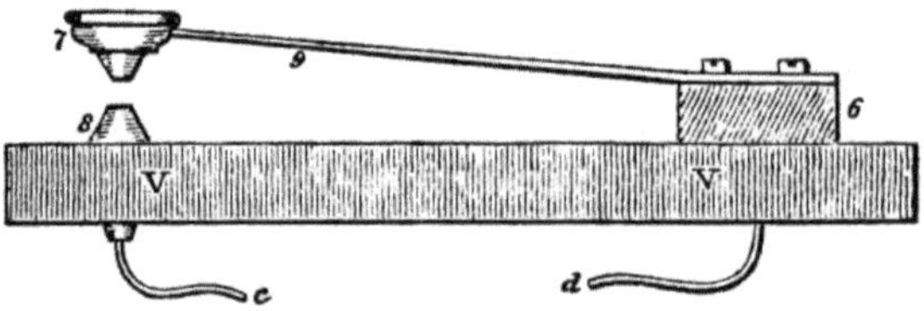

The key or correspondent is represented by 6, 7, 8, 9. Another view of it is more distinctly seen in figure 11. The same letters in each, represent the same thing. V and V is the platform. 8 is a metallic anvil, with its smaller end appearing below, to which is soldered the copper wire *c*. 7 is the metallic hammer, attached to a brass spring, 9, which is secured to a block, 6, and the whole to the platform, V V, by screws. A copper wire passes through the whole, and is soldered to the brass spring at 6. The key or correspondent is used for writing upon the register at the distant station, and both it and the register are usually upon the same table.

Having now explained the Register, Key and Battery, we proceed to describe the arrangement of the conductors or wires connecting distant stations, and the mode by which the earth, also, is made a conductor of this subtle fluid.

The term *circuit* used frequently in this work, has reference to the wire, which, commencing at the positive pole of the battery, goes to any distance and returns to the negative pole of the battery. When its going and returning are continuous or unbroken, the circuit is said to be *closed* or *complete.* When it is interrupted, or the wire is disconnected, the circuit is said to be *broken* or *open.*

When a magnet or key or battery is spoken of as being *in the circuit*, it has reference to the use of the wire belonging to the key, magnet or battery, respectively, as a part of the circuit.

There are three modes of arranging the wires, so as to communicate between two distant stations. Two of these modes are *inferior*, as they furnish but one circuit for the termini, and consequently obliging one station to wait, when the other is transmitting, both stations not being able to telegraph at the same time. These two modes are called the *dependent circuits.* The first mode is, where two wires are used, of which figure 12 is a diagram. B represents Baltimore, and W Washington; *m* is the magnet or register; *k* the key, and *bat* the battery, all at the Baltimore station; *m'* is the magnet or register; *k'* the key at the Washington station. The lines, represent the wires upon the poles, connecting the two stations, and are called the east and west wires. In this arrangement of the wires and also in the second, the key (which has been explained in a preceding figure, 11, and shown at 6 and 7 to be open) must be closed at both stations, in order to complete the circuit, except at the time when a communication is being transmitted.[6] For the purpose of closing the circuit at the key, a metallic wedge is used, which is put in between the anvil 8 and the hammer 7, and establishes the circuit. Supposing the battery is in action, and B has a communication for W: he opens his key, by removing the wedge, and sends his message. The galvanic fluid leaves the point, P, of the battery, and goes to *k*, to *m*, along the east wire to *k'*, to *m'*, and back by the west wire to N pole of the battery. In the same manner it proceeds along the wires, if W is writing to B. In this arrangement, the direction of the galvanic current is the same, whether B or W is communicating, unless the poles of the battery are reversed.

Fig. 12.

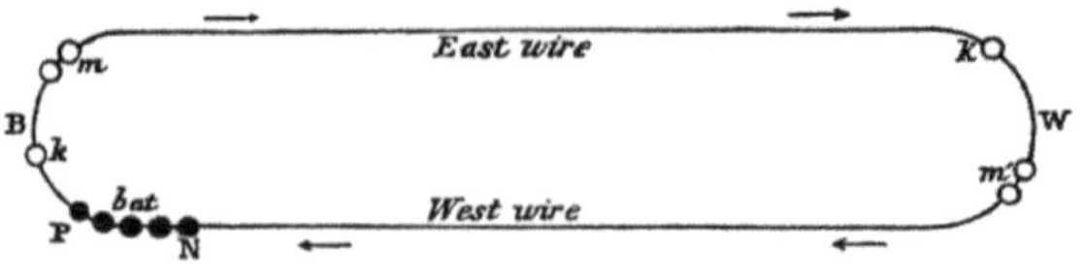

Fig. 13.

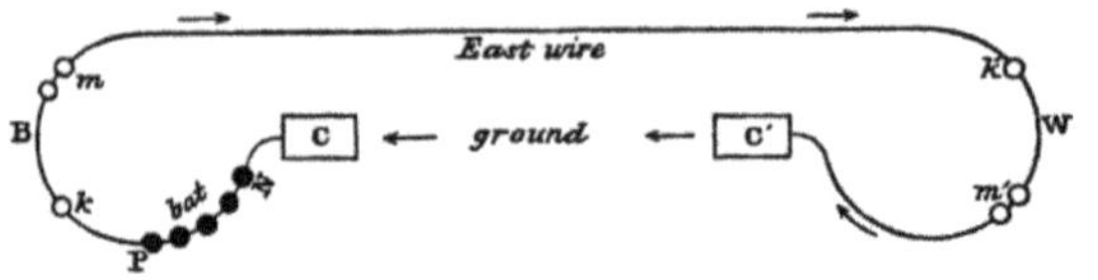

The second mode has but one wire and the ground, represented by figure 13. The use of the ground as a conductor of the galvanic fluid, between two distant points, is to many a mystery. But of the fact there is no question. The above diagram exhibits the manner in which the east wire and ground were used from the first operation of the Telegraph, until the close of the session of Congress, June, 1844. In this diagram, we will minutely follow the course of the galvanic current. B represents Baltimore, and W Washington; C represents a sheet of copper, five feet long and two and a half feet wide, to which a wire is soldered and connects with the N pole of the battery. This sheet of copper lies in the water at the bottom of the dock, near the depot of the Baltimore and Ohio Rail Road, Pratt street. From P of the battery, the wire proceeds to *k*, the key, then to *m*, the magnet or register, then it is the *east wire* to *k'*, the key at W, then to *m'*, the magnet or register, then to the *copper sheet*, C', buried beneath the brick pavement in the dry dust of the cellar of the capitol. The direction of the current is from P of the battery to *k*, to *m*, and along the east wire to *k'*, to *m'*, and to C', where it is lost in the earth; but reappears at the copper plate, C, at B, and thence to the N pole of the battery, having completed its circuit. It is, therefore, certain, that one-half of the circuit is through the earth. From B to W the east wire is the conductor; and from W to B the ground is the conductor. In this arrangement, the west wire is thrown out, and is no part of the circuit; while the earth has been made a substitute for it.

Fig. 14.

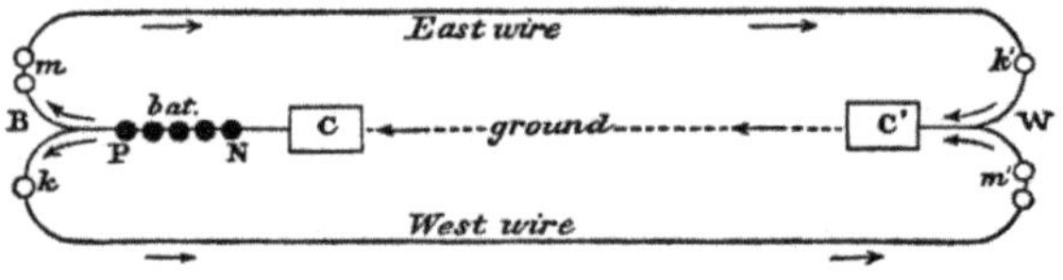

The last diagram, as has been stated, exhibits the plan of the wire and ground, as used for telegraphic purposes, from its first operation, until the adjournment of Congress in 1844, being prevented from completing the arrangement of the third mode from the throng of visitors, that pressed to see its operation. After the close of the session, the following arrangement of the wires was made, as shown in the diagram, figure 14, by means of which, both stations could transmit at the same time, with one battery for both, and the keys were not required to be closed. It is called the two *independent circuits*. Here the west wire is used for transmitting from B to W; and the east wire from W to B. The copper plates at B and W remain as they are described in the second plan. *Bat*, the battery, at B is used in common for both circuits. It is simply necessary here to designate the course which the fluid takes when both lines are in operation, viz. B transmitting to W; and W to B. In the former case, the current is from P of the battery to *k*, then the *west wire*, then to *m'*, at W, then to *C'*, thence through the ground to *C* at B, and then to the N, or negative pole of the battery, as shown by the arrows. In the latter case, the current is from P of the battery to *m*, then the *east wire*, then to *k'*, at W, thence to *C'*, thence through the ground to C at B, thence to the N, or north pole of the battery, as shown by the arrows. This arrangement, by which one battery is made efficient for both circuits at the same time, where two were formerly used, was devised by Mr. Vail, assistant superintendent, in the spring of 1844, and has contributed much to diminish the care and expense in maintaining that part of the apparatus of the telegraph. One battery being now used instead of two. By the above diagram, it will be perceived that the *ground* is common to both circuits, as well as the *battery*, and also the wire from the N pole of the battery, to the copper plate, C; and from the copper plate, C', to the junction of the two wires near the two arrows. For the purposes of telegraphic communication they answer as well as though there were four wires and two batteries. Instead of using the ground between C and C', a wire might be substituted, extending from the N pole of the battery to the junction of the wires at the two arrows at W. The arrangement of the wires, battery, keys, magnets or registers at both stations, with the ground, as shown in figure 14, is the plan now used for telegraphic operations between B and W; and has many decided advantages over the arrangements of figures 13 and 14. First. In both of those arrangements, the circuit is obliged to be kept closed, when neither station is at work; and as the battery is only in action when the circuit is closed, it follows that the battery will not keep in action as long as when the circuit is allowed to remain open, as in the use of the third plan, figure 15. Second. There is an advantage in dispensing with the use of the metallic wedge, which is liable to be forgotten by the operator. Third. The attendant may occasionally leave the room, and is not required to be in constant waiting, as the clock work is put in motion and stopped by the operator at the other end, and the message written without his presence. But in the first and second arrangement, the apparatus for putting in motion and stopping the clock work, is entirely useless. The attendant being obliged to put it in motion and stop it himself.

We will now proceed to describe the modus operandi of transmitting intelligence from one station to another; the arrangement being as in figure 14; *k* is the key of the operator at Baltimore, and *m'* represents his register, or writing desk, at Washington; *k'* is the key of the operator at Washington, and *m* his register, or writing desk, at Baltimore. Each has the entire control of his respective register, excepting, only, that each operator winds up the other's instrument, and keeps it supplied with paper. It will also be borne in mind, that each circuit is complete, and everywhere continuous, except at the keys, which are open. If, then, the hammer is brought in sudden contact with the anvil, and permitted as quickly as possible to break its contact by the action of the spring, and resume its former position, the galvanic fluid, generated at the battery, flies its round upon the circuit, no matter how quick that contact has been made and broken. It has made the iron of the electro magnet a magnet; which has attracted to it the armature of the pen lever; the pen lever, by its steel pen points, has indented the paper, and the pen lever has, also, by the connecting wire with the break; taken it from the friction wheel; this has released the clock work, which, through the agency of the weight, has commenced running, and the two rollers have supplied the pen with paper. But, as only one touch of the key has been made, the clock work soon stops again, if no other touches are made, by the action of the break upon the friction wheel.

This shows the whole operation of the Telegraph, in making a single dot by a single touch of the key. In order now to explain more fully the operation of the steel pen points upon the paper, which is in contact with the grooved roller, let there be made four touches at the key; this will be sufficient to start the clock work, and allow the paper to have attained a uniform rate; then let six touches be made at the key. The contact has been made six times and broken six times. Each time it is closed, the electro magnet, as heretofore explained, attracts to it, with considerable force, the armature of the pen lever, carrying up the steel pen points against the paper, 2, under the steel roller, S. The three points of the pen, falling into the three corresponding grooves of the roller, carry the paper with them and indent it,[7] at each contact. There then appear upon the paper, as it passes out from under the rollers, six indentations, as if it had been pressed upon by a blunted point, such as the end of a knitting needle would be supposed to make, when pressed upon paper, placed over a shallow hole, but in such a manner as not to pass through the paper, but raising the surface, as in the printing for the blind. These indentations of the paper are the marking of the pen, but varied in the manner now to be described.

By examining the telegraphic alphabet, the characters will be found to be made up of dots: short and long lines—and short and long spaces. A single touch of the key, answers to a single dot on the paper of the register; which represents the letter, E. One touch of the key prolonged, that is, the contact at the key continued for about the time required to make two dots, produces a short line, and represents T. A single touch for about the time required to make four dots, is a long line, and represents L. A single touch for about the time required to make six dots, is a still longer line and represents the 0 of the numerals. If the use of the key be suspended for about the time required to make three dots, it is a short space, used between letters. If suspended for the time required to make six dots, it is a long space, used between words, and a longer space is that used between sentences. These are the elements which enter into the construction of the telegraphic characters, as used in transmitting intelligence. The alphabet is represented by the following combination of these elements.

ALPHABET.

·_	_···	·· ·	_··	·	·_·	__·	····	··	_·_·	_·_
A	B	C	D	E	F	G	H	I	J	K
—	__	_·	· ·	·····	··_·	· ··	···	_	··_	···_
L	M	N	O	P	Q	R	S	T	U	V
·__	·_··	·· ··	··· ·	· ···	·__·	··_··	···_·	····_		
W	X	Y	Z	&	1	2	3	4		
___	······	__··	_····	_··_	———					
5	6	7	8	9	0					

Suppose the following sentence is to be transmitted from Washington to Baltimore:

```
-  ....  .     ._  --  .  ...  ..  ...  ._  -.     .  —  .  ...
T   h    e     A   m   e   r   i    c   a   n      E  l  e   c

-  ...  ..     --  ._  --.  -.  .  -  ..  ...     -  .  —  .
t   r   o      M   a   g    n   e  t  i   c       T  e  l  e

--.  ...  ._  .....  ....     ..  -.  ..._  .  -.  -  .  -..     _...  ....
g    r    a   p      h        i   n   v     e  n   t  e  d       b     y

.....  ...  ..  ._.  .  ...  ...  ..  ...     ...  ._.  _...     --  ..
P      r    o   f    e  s    s    o   r       S    F    B        M   o

...  ...  .     ..  ._.     -.  .  .--     ... ..  ..  ...  -.-
r    s    e     o   f       N   e  w       Y       o   r    k

..  -.     _...  ..  ._  ...  _..     ..  ._.     -  ....  .
o   n      b     o   a   r    d       o   f       t  h     e

.....  ._  ....  -.-  .  -     ...  ....  ..  .....     ...  .._  —
p      a   c     k    e  t     s    h     i   p         S    u    l

—  .. ..     ...  ._  .....  -     .....  .  —  —     ..  -.
l  y         C    a   p      t     P      e  l  l     o   n

....  .  ...     .....  ._  ...  ...  ._  --.  .     ._.  ...  ..  --
h     e  r       p      a   s    s    a   g    e     f    r    o   m

....  ._  ..._  ...  .     -  ..     -.  .  .--     ... ..  ..  ...  -.-
H     a   v     r    e     t  o      N   e  w       Y       o   r    k

..  ...  -  ..  _...  .  ...     .--.  -....  ...-.  ..-..
O   c    t  o   b     e  r       1     8      3      2
```

It is evident, as the attendant at Baltimore has no agency in the transmission of this message from Washington, his presence, even, is not absolutely required in the telegraph room at Baltimore, nor is it necessary, previously, to ask the question, *are you there*? The operator at Washington transmits it to Baltimore, whether the attendant is there or not, and the telegraphic characters are distinctly recorded upon the paper of the Baltimore-register. If he omits a letter at the key, in Washington, it is omitted on the paper in Baltimore. If he has added at the key in Washington, it is also upon the paper in Baltimore, nothing more or less is marked upon it.

Specimen of the Telegraphic Language.

- ·· _· _ ···· · ·· ·· · ·_ · ·· ·__· _···· ···_·
- ··_·· · · _· __ ·· ·· ···_ · · ·· ·· ·_ __· · ····
- · · __ ·· ·_· · ·· · · __ · ··_ · ·· · · ····· · _
- ···· · · ___ · ·· · _ · ·· ·· ·· · ·_ ___ · ·_··
- ····· · · ·· ·· __ · _· _ · · ·_· ·_· · ·· ·_ _· _·_
- ___ ·· _· ··_ ····· · · _· ·_ ·__ ·· · ·· · ···
- · · __ · ····_ __ ·· ___ · ··· ·· _· ___ · _· __·
- _ ···· ·__ ·_ ··· ·· · ·_ ··· ··_ ·_ ___ ___ ·· ··
- · ·· · ·· · ·_ ___ ___ · _·· _ · · __ ·· ·· __ ·
- _· _·· ·· _· ·_ ·· · · · _· ···_ · · ·· ··· ·_ _
- ·· · · _· ·__ ·· _ ···· ·__· · · ·_· _ ···· ·
- ····· ·_ ··· ··· · _· __· · · ·· ··· ·· _· ·__ ····
- ·· ·· · ···· · ·_·· ····· · · ·· ·· __ · _· _ ·· _
- ·__ ·_ ··· ·_ ··· ·· · · · ·· _ ·_ ·· _· · _·· _
- ···· ·_ _ _ ···· · · ___ · ·· · _ · ·· ·· ·· · ··
- _ ·· ·· _ · ·· ·_ ···_ · ___ ___ · _·· _ ···· · ··
- · · ··_ __· ···· _ ···· · ·__ ···· · · ___ · ·· ·
- ·· · ·· ·· · ··_ ·· _ ·· _· ·_ _ ·· __ · _·
- · · _ ·_ ····· ····· · ·· · ·· · ·· ·_ _··· ___ · _···
- ··_ _ ·_ ····· ····· ·_ · ·· · _· _ ___ ·· ·· ·· _·
- ··· _ ·_ _· _ ·_ _· · · · ··_ ··· ·· _ ·· __ __ ·

- _·· ·· ·_ _ · ___ ·· ·· · · ·· · ·· · ··_ · ·· · ·· ·
- _·· _ · · __ · _ ···· ·_ _ ·· ·_· _ ···· ·
- ····· · ·· · ··· · _· ·· · · · · ·_· · ___ · ·· · _
- · ·· ·· ·· · ·· _ ·· ·· ·· · · · ··_ ___ _·· _··· ·
- __ ·_ _·· · ···_ ·· ··· ·· _··· ___ · ·· _· ·_ _·
- ·· ·· _·· · ··· ·· · ·· · _·· ····· ·_ · ·· _ · ·
- ·_· _ ···· ·· ··· ·· · ·· · ·· ·· · ··_ ·· _ ·· _
- ·__ · · ··_ ___ _·· _· · · _ _··· · _·· ·· ·_· ·_·
- ·· ·· · ··_ ___ _ _ · · ·· · · · _· ··· _ · ·· ··_
- ·· · _ ·_ ··· ·· ·· ··· _ · __ · · ·_· ··· ··
- __· _· ··· _··· ·· ·· ·__ ···· ·· ·· · ···· ·· _·
- _ · ___ ___ ·· __· · _· ·· · · ·· · · · ··_ ___ _··
- _··· · ·· _· ··· _ ·_ _· _ ·_ _· · · · ··_ ··· ___
- ·· ·· _ · ·· ·_ _· ··· __ ·· _ _ · _·· _ ···· ·
- _ ···· · · ··_ __· ···· _ _ ···· ··_ ··· ·· · · · _·
- ·· · · ·· ···_ · _·· _ · · · · _·_ ··· _ · ·· · ·
- _· __· ···· · · ___ _·· · · ·_· __ ·· ·· __ ··
- _· _·· ·· _· _ ···· · ___ · ·· ··· ··_ · ··· ·__
- ···· ·· ·· · ···· _ ···· · ···_ · · ·· ·· ·_ __· ·
- ·_ ·_· ·_· · · · ·· _·· · _·· · ··· ·· ····· ___
- ·_ _· _· · _·· ·_ ··· ·· ·· ··· _ · __ · · ·_·
- ··· ·· __· _· ··· · ··· ·_ _· ·_ ····· ····· ·_ · ··
- ·_ _ ··_ ··· _ · · ·· · ·_ · ·· · ·· ·· ·· ·· _ ··
- _· _ · · · ·_· ·_· · ·· · _ ·· ·· · ·_ ··· _ ·_
- ··· ····· · ·· · ·· · ··· · · ·_· _ ·· ·· ····· ·
- ·__ ···· ·· ·· · ···· ·· ···· ·_ _·· _·· · ···_ ··
- ··· · _·· ·_· · · · ·· _ ···· ·· ··· ····· ··_
- · ·· ····· · · ··· · _ ···· · ·_· ·· · ·· ··· _

- ·__ · · _·_ ·_ ·_· _ · · ·· __ ·· ·· ·_ · ·· · ··
- ·· ···_ ·_ ____ ···· · · __ · · ··· ·_ ____ _ ····
- · · ··_ __· ···· _ ···· · · ·· · ··· _ · · ·_· _
- ···· · __ ·_ ·· · ···· ·· _· · · ·· ·· ·· ·__ ·_ ···
- ····· ____ ·_ _· _· · _·· ·· ·· · _ ·_· · ·· · · __
- _ ···· · ····· · ·· · ··· ··· ··_ · ·· · · · ·_· ··_
- _· ·_ ···_ · · ·· _·· ·_ _··· ____ · _·· ··_ _ ·· ·
- ··· ·· ·__ ·_ ··· ·· · · · __ ····· · ____ ____ ·
- _·· _ · · ····· · · ··· _ ····· · · _· · __ ·· ··
- · ·_·· ····· · · ·· ·· __ · _· _ ··· · ··· ·__ ·_
- ··· _· · · _ ·_ _··· ____ · _ · · _ · ··· _ _
- ···· · ·__ ···· · · ____ · ····· ____ ·_ _· ··_ _·
- _ ·· ____ ·__ ·· _ ···· ·· _· ·_ ·_· · ·__ ·__ · ·
- _·_ ···
- ··· ·_· _··· __ · · · ·· ··· ·

From the peculiarity of the motion obtained at the pen lever by the action of the battery upon the electro magnet, it is evident that a few elements only are presented upon which to base the telegraphic characters. The motion of the lever, to which is attached the steel pen points, is vibratory; but capable of being so controlled as to cause it to retain either of its positions (that is, up or down) as long, and at such intervals, and in as quick succession as the operator may choose. Therefore, every sort of combination which dots, lines and spaces, in any succession, and of any length can make, are here as much at the pleasure of the telegraphic manipulator, as the English alphabet is with the letter writer. So that if from this countless variety, twenty-six of the most simple, to represent letters, and ten to represent the numerals, shall be taken, we come at once into possession of the means of representing words and sentences, by new, but intelligible characters, and through them, can be conveyed as clearly, and as concisely, as if they were given viva voce, or written in Roman characters. Such is the alphabet given above. This conventional alphabet was originated on board the packet Sully, by Prof. Morse, the very first elements of the invention, and arose from the necessity of the case; the motion produced by the magnet being limited to a single action.

During the period of thirteen years, many plans have been devised by the inventor to bring the telegraphic alphabet to its simplest form. The plan of using the common letters of the alphabet, twenty-six in number, with twenty-six wires, one wire to each letter, has received its due share of his time and thought. Other modes of using the common letters of the alphabet, with a single wire, has also been under his consideration. Plans of using two, three, four, five and six wires to one registering machine, have, in their turn, received proportionate study and deliberation. But these, and many other plans, after much care and many experiments, have been discarded; he being satisfied that they do not possess that essential element, *simplicity*, which belongs to his original first thought, and the one which he has adopted. A detailed account of these various plans with fewer or more wires, might be given here, but it will suffice merely to present the alphabet adapted to a register, using 2, 3, 4, 5, or 6 wires, with a separate pen to each wire capable of working together, or in any succession. It is obvious that every additional pen will give an additional element to increase the combination, and were there any real advantage in such an arrangement it would have been adopted long since.

No. 1.

Alphabet for two pens, operating together or in succession.

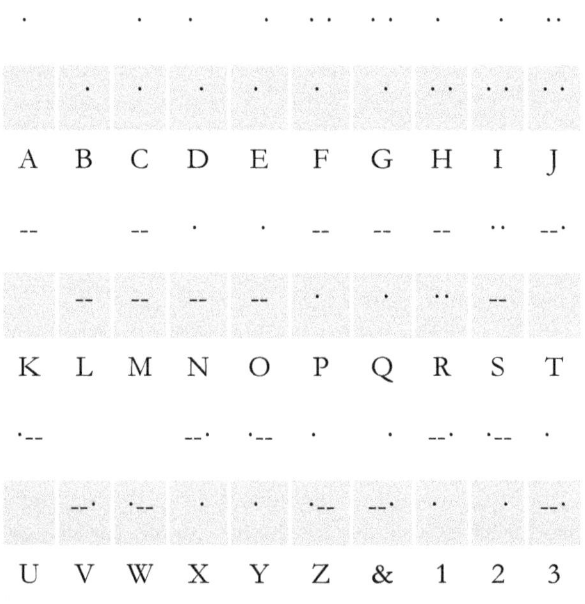

·	·--	--·	--·	·--	--	--			
·--	·--	--·	--	--	--·	·--			
4	5	6	7	8	9	0			

No. 2.

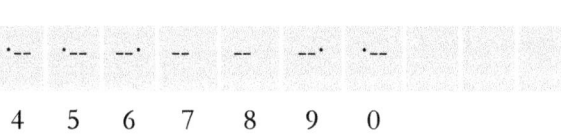

Alphabet for three pens, operating together or in succession.

·			·		·	·	· ·		
	·		·	·	·			· ·	
		·		·	·	·			· ·
A	B	C	D	E	F	G	H	I	J
· ·			· ·		·	·		·	·
·	· ·	· ·	·	·	· ·	· ·	·	· ·	· ·
	·	·		· ·			· ·	·	·
K	L	M	N	O	P	Q	R	S	T
· ·		· ·	· ·	· ·	·	·	·	·	· ·
· ·	· ·						·	·	·
	· ·	· ·	·		· ·	· ·	· ·	· ·	·
U	V	W	X	Y	Z	&	1	2	3
· ·	· ·	· ·	· ·	· ·	·	·			
·	·	·	· ·	· ·	· ·	· ·			
·	· ·	· ·	·	·	· ·	· ·			

4 5 6 7 8 9 0

No. 3.

Alphabet for four pens, operating together or in succession.

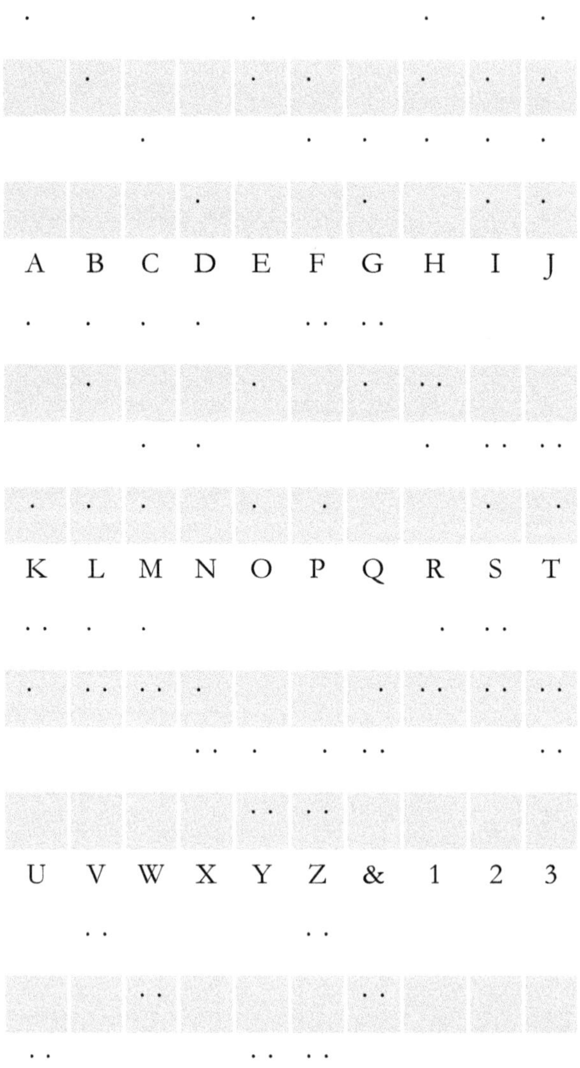

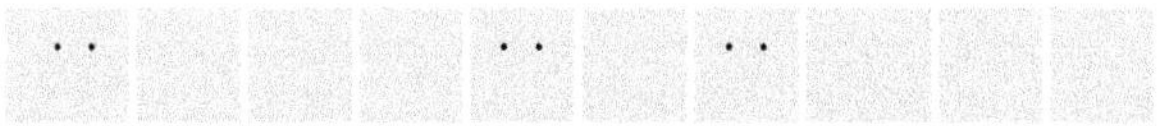

4 5 6 7 8 9 0

No. 4.

Alphabet for five pens, operating together or in succession.

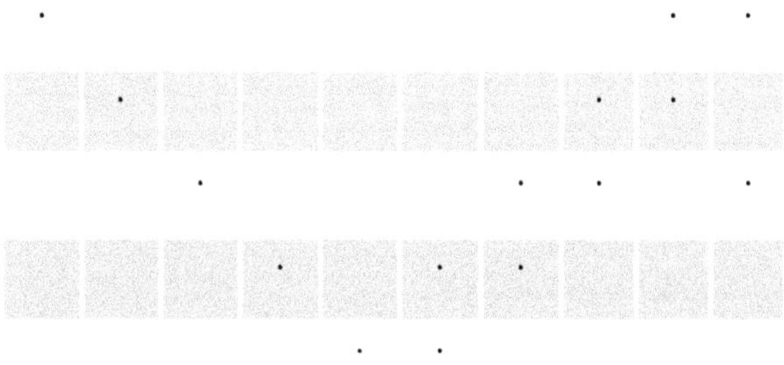

A B C D E F G H I J

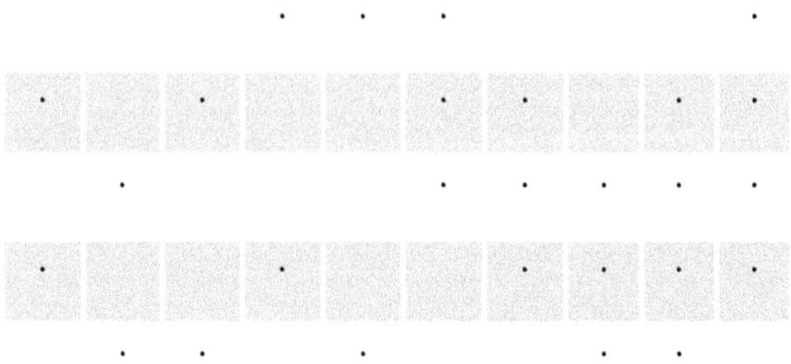

K L M N O P Q R S T

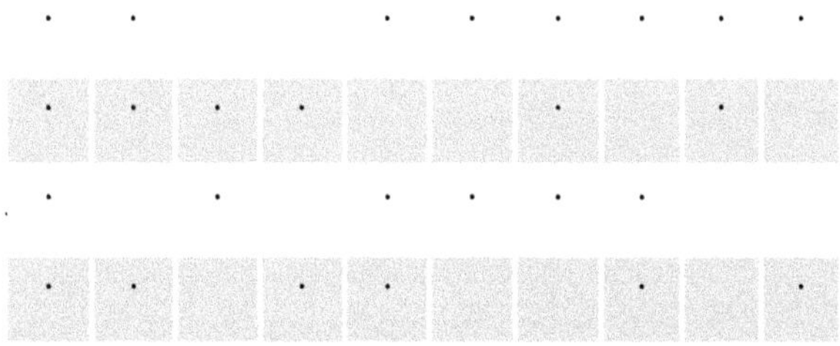

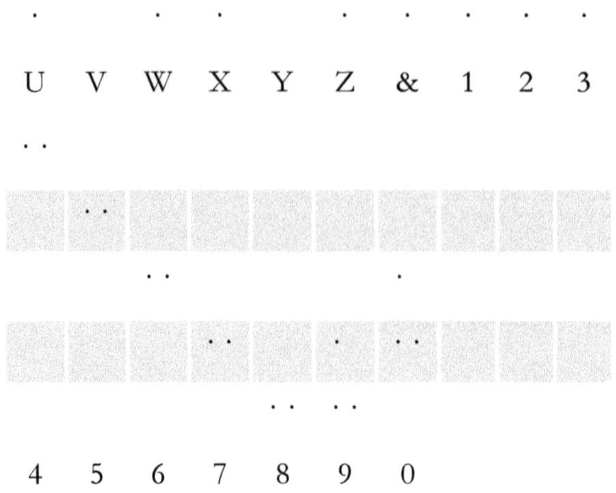

No. 5.

Alphabet for six pens, operating together or in succession.

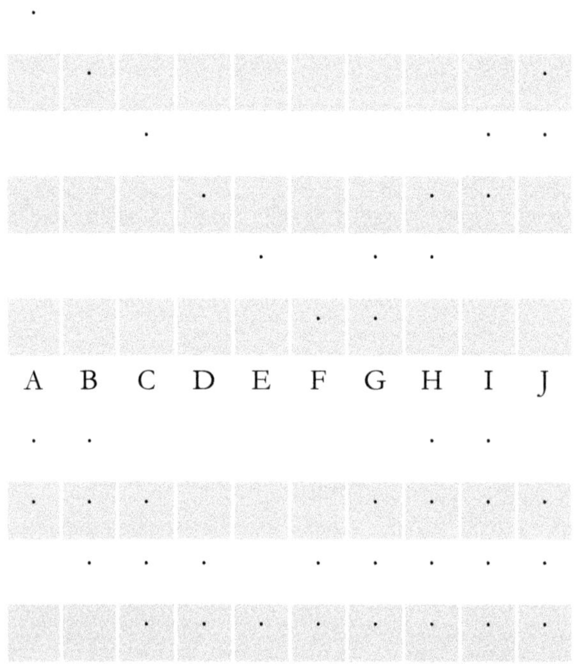

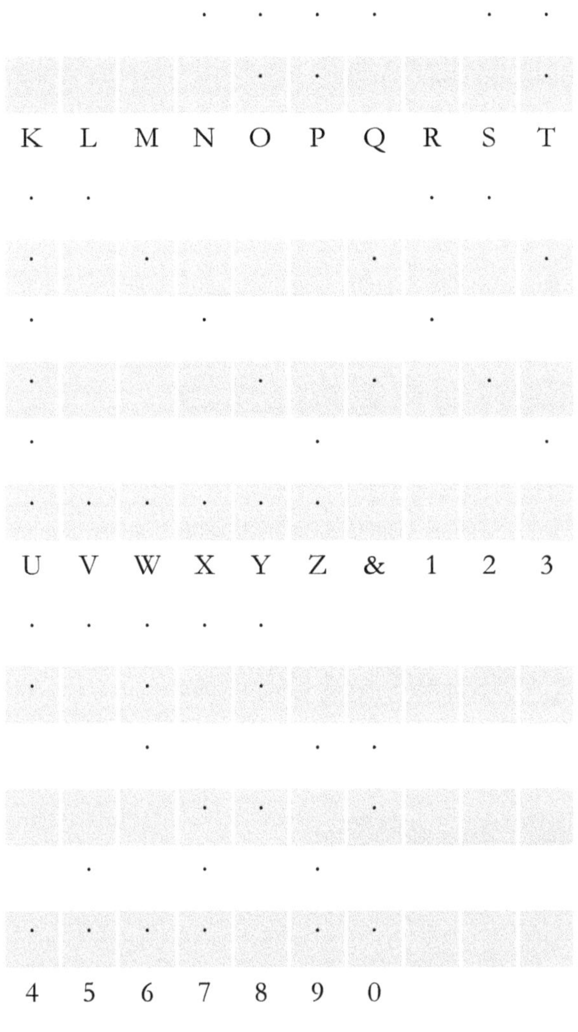

CORRESPONDENT OR KEY.

The modes of manipulation for sending intelligence, which at various times have been invented by Prof. Morse, are more various than any other part of the machinery of the telegraph. A few of them will now be described. The first method, invented as early as the year 1832, was that of using a type, resembling saw-teeth, set up in long frames, and made to pass under a lever, by means of machinery, at a uniform rate, for the purpose of closing and breaking the circuit, in a manner hereafter to be described. The following figure, 15, represents the saw-teeth type. The top of the narrow tooth corresponds with the *dots* of the letters, and the long tooth, with the *lines* of the letters. For instance, A, has one tooth for a dot, and a long tooth for a line, which is the telegraphic letter A; then follows a space at the end of the type, corresponding with the short space between two letters.

Fig. 15.

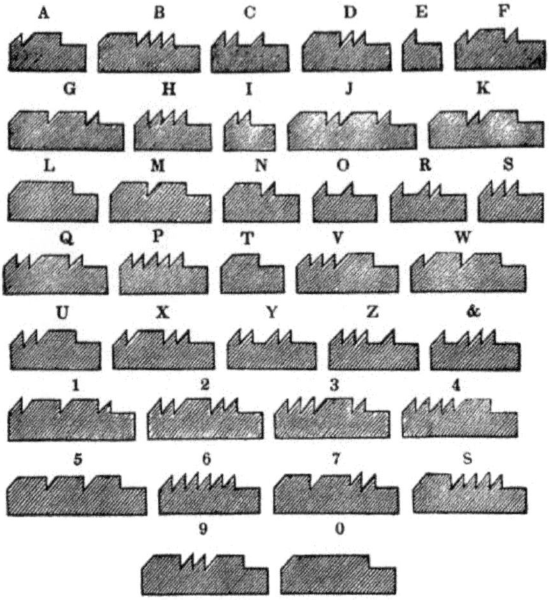

Fig. 16.

These type were set up in a cavity, made by putting two pieces of long rules of brass plate together, side by side, with a strip of half their width between them; so as to make the cavity sufficiently large to receive the type. This was denominated the *port rule*, and is represented in figure 16 by A A. Parts of the type are seen rising above the edge of the *rule*, and below it are seen the cogs, by which, with the wheel, V, the pinion, L, and the crank, O, the port rule, with its type, were carried along at an uniform rate in a groove of the frame, K, R, under the short lever, C, which has a tooth, or cam, at its extremity. J is a support, one on each side of the frame, for the axis of the lever, B and C, at its axis, I; *a* and *i* are two brass or copper mercury cups, fastened to the frame. These cups have the negative and positive wires soldered to them, N and P. D and H are the ends of *one* copper wire, bent at right angles at that portion of it fastened to the lever, B. The ends of the copper wire are amalgamated, and so adjusted, that when the lever is raised at C, by the action of its cam, passing over the teeth of the type, the lever, B, is depressed, and the wires, D and H, dip into the mercury cups, and thus complete the connection. This plan worked well, but was too inconvenient and unwieldy.

The second method was upon the same principle, with a more compact arrangement. The type being put into a hopper and carried one by one upon the periphery of a wheel, the teeth acting upon a lever in the same manner as in the figure preceding. The wheel being horizontal.

Fig. 17.

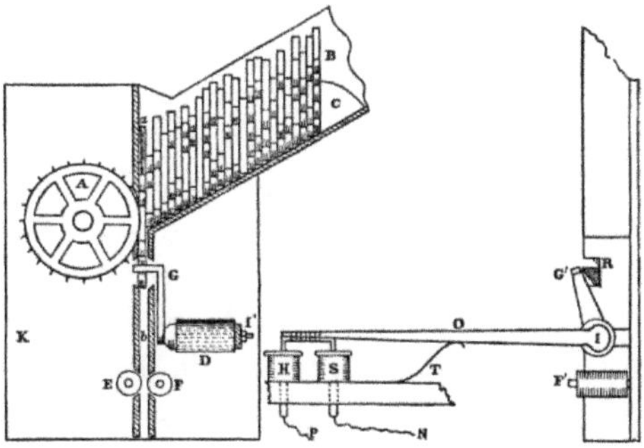

The third plan differed only in one respect, instead of the types moving in a circle, they were made to move in a straight line. Figure 17 represents that instrument. The type were all made with small holes through their sides, so as to correspond with the teeth of the wheel, A, driven by clock work and weight. K is the side of the frame containing the clock work. B is the hopper containing the types, with their teeth outward. The hopper is inclined at an angle, so that the type may slide down as fast as one is carried through the cavity, *a* and *b*. C is a brass block to keep the type upright, and sliding down with them. E and F are two small rollers, with springs (not shown) to sustain the type, after the wheel, A, has carried them beyond its reach. G is a lever for the same purpose as C in figure 16. D its support, through which its axis passes. I' is the long lever, O, of the left side figure, to the end of which, is the bent wire in the mercury cups, H and S, and to which are soldered the wires, P and N. T is the spring to carry back the lever, O. F' is one of the small rollers, and G' the short lever. At R may be seen a part of one of the type passing; the tooth having the short lever upon its point, thereby connecting the circuit at the mercury cups, H and S, by the depression of the long lever, O. The hopper, B, may be of considerable length, and at a less angle. When a communication is to be sent, it is set up in type, and put in the hopper. The clock work is then put in motion, and the wheel, A, will carry them down one by one. In this manner, the cam on the end of the lever, G, will pass over all the teeth of the type, as in the plan shown by figure 16.

The fourth plan was by means of keys, one for each letter and numeral. By pressing upon any one of the keys, it wound up the clock work of the instrument. The key being instantly released, and returning gradually to its former position, produced the closing and breaking of the circuit required to write its character upon the register.

The fifth plan is in some respects similar to the last, but much more simple, and requiring less time in transmitting intelligence. Figure 18 exhibits a view of the keyed correspondent, with its clock work. A' represents a top view of it, and B' is a side or front view. 1 1 1 1, of both views, represent the long cylinders of sheet brass, covered with wood or some insulating substance, except at the black lines, which represent the form of the letters, made of brass, appearing at the surface of the cylinder and extending down and soldered to the interior brass cylinder. A cross section of the cylinder is seen at D', of which the blank ring is the brass cylinder, and the blank openings to the outer circle the metallic forms of the letter J, and the shaded portion of the circle represents the insulating substance, covering the whole surface of the cylinder, except, where the letter-forms project from the interior. It is obvious that every letter and parts of each letter are in metallic connection with the brass cylinder. At each end of the cylinder is a brass head, with its metallic journal, and the journal or arbor turns upon its centre in a brass standard, 17, secured to the vertical frame. To this standard is soldered the copper wire, N, connected with the negative pole of the battery. There are together thirty-seven letters and numerals upon the cylinder, and made to correspond to the letters of the telegraphic alphabet. To each of these, there is a separate key, directly over the letter cylinder. Each key has its button, with its letter, A, B, C, D, &c., marked upon it, and beneath the button in a frame of brass, is a little friction roller. The key is a slip of thin brass, so as to give it the elasticity of a spring, and is secured at the thicker end by two screws to a brass plate, extending the whole length of the cylinder, so as to embrace the whole number of keys. This plate is also fastened to the vertical mahogany frame. At the right hand end of the brass plate is soldered a copper wire, leading to the positive pole of the battery, after having made its required circuit through the coils of the magnet, &c. It is clear, that if any one of the keys is pressed down upon any portion of a metallic letter, that the circuit is completed; the galvanic fluid will pass to the brass plate to which, P, wire is soldered; thence along the plate to the spring or key; then to the small friction roller beneath the button; then to that portion of any letter with which it is in contact; then to the interior brass cylinder, to the arbor; then to the brass standard, and along the negative wire, soldered to it, to the battery. We have now to explain in what manner, the cylinder is made to revolve, at the instant any particular key is pressed, so that the metallic form of the letter may pass at an uniform rate under the roller of the key; breaking and connecting the circuit so as to write at the register, with mechanical accuracy, the letter intended.

Fig. 18.

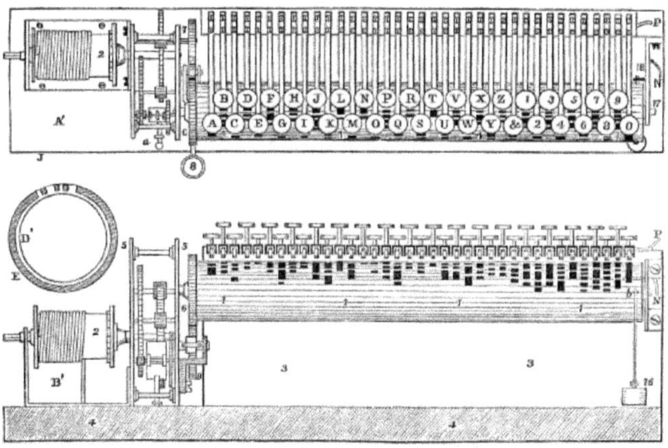

Fig. 19.

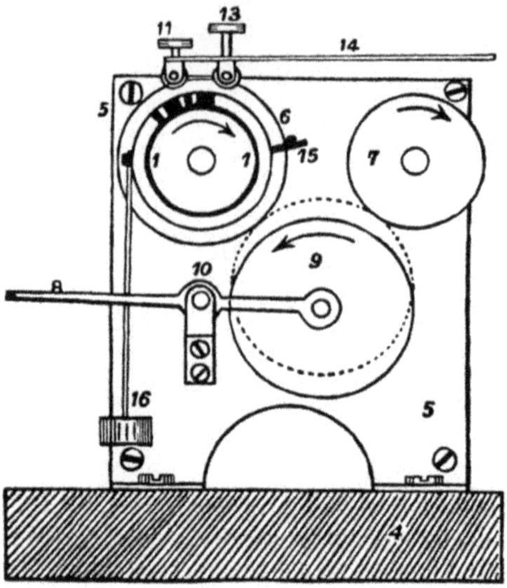

4 4 is the platform upon which the parts of the instrument are fastened. 3 3 is the vertical wooden back, or support, for the keys and brass standard, 17. 2 is the barrel of the clock work contained within the frames, 5 5. With the clock work, a fly is connected for regulating its motion, and a stop, a, for holding the fly, when the instrument is not in use; 6 is a very fine tooth wheel, on the end of the letter cylinder; 7 is also a fine tooth wheel, on a shaft driven by the clock train. In the front view is seen, at 9, another fine tooth wheel, suspended upon a lever, the end of which lever is seen at 8, figure 18, A'. 18 is a stop, in the standard, 17, to limit the return motion of the cylinder, which also has a pin at 18, at right angles with the former. 16 is a small weight, attached to a cord, and at its other end, is fastened to the cylinder at *b*. The relative position of the three fine tooth wheels, and the lever, 8, are better seen in a section of the instrument, figure 19. The same figures represent the same wheels as in the other views, A' and B'. 7 is the wheel driven by the weight and train. 6 the wheel, on the end of the cylinder, to which motion is to be communicated, and 9 is the wheel, suspended upon the end of the lever, 8, of which 10 is its centre. 1 1, is the brass lettered cylinder. 11 and 13 the buttons of the two keys, one a little in advance of the other. 14 is the spring and the two friction rollers of the key, may be seen directly under the buttons. 15 is the stop pin. 16 the small weight and cord attached to the cylinder, to bring it back after each operation. 4 4 is the end view of the mahogany platform. The arrows show the direction which the wheels take, when the lever is pressed with the thumb of the left hand at 8, so as to bring wheel 9, up against 7 and 6, connecting the two, as shown by the dotted lines. Wheel, 7, communicating its motion to 9, and 9 to 6, which causes the

metallic letters to pass under the rollers in the direction of the arrow. Now, in order to use the instrument, let it be supposed a letter is to be sent. The stop, *a*, figure 18, A', is removed from the fly, and the clock work is set in motion by the large weight. Then the thumb of the left hand presses upon the *lever*, 8, at the same time, *key*, R, is pressed down by the finger of the right hand, so that the small roller comes in contact with the cylinder. At the instant the roller touches the cylinder, the letter begins to move under the small roller, making and breaking the circuit with mechanical accuracy. When the letter has passed under the small roller, the thumb is taken off the lever, 8, and the finger from the key, R. The cylinder is then detached from its gear wheel, 9, and the weight, 16, instantly carries it back to its former position, in readiness for the next letter. Then the *lever*, 8, and the *key*, *E*, are pressed down at the same instant for the next letter, and it is carried under the small roller in the same manner as the first, which, when finished, the wheel, 9, is suffered to fall, and the cylinder returns to its natural position again. The same manipulation is repeated for the remaining letters of the word.

In the following figure, 20, is represented the flat correspondent. It somewhat resembles the keyed correspondent, but without keys or clock work. A represents the arrangement of the letters, presenting a flat surface. Those portions in the figure, marked by black lines and dots, represent the letters which are made of brass. That portion which is blank, represents ivory or some hard insulating substance, surrounding the metal of the letters. As in the keyed correspondent, each letter and parts of each letter extend below the ivory and are soldered to a brass plate, the size of the whole figure, A. A sectional view of this is seen at 1 1, which is ivory, and 2 2, the brass plate below. The whole is fastened to a table, B. 5' and 5' is a brass plate, called the guide plate, with long openings, represented by the blanks, so that when the guide plate, 5' 5', is put over the form, A, each opening is directly over its appropriate letter, and is a little longer than the length of the letter. 4' and 4' is the wooden frame, to which the guide plate is secured. The ends of this frame are seen in the sectional figure at 4 4, and the guide plate at 5 5. The dark portions of which, represent the partitions, and the blanks the openings. It will be observed here that the plate, 5 5, resting upon the wooden frame 4 4, is completely insulated from the brass letter plate 1 1, and 2 2. The blank space between them showing the separation. It is, however, necessary that the guide plate should be connected with one pole of the battery, and the letter plate with the other pole. For this purpose a brass screw, F, passes up through the table, B, and through 4, into the guide plate 5 5. The head of the screw has a small hole through it, for passing in the end of the copper wire, G, from the battery, and a tightening screw below, by which a perfect connection is made. At D, is another screw, passing through the table, and into the letter plate, 2 2. To the head of this screw is also connected another copper wire, E, extending to one of the poles of the battery.

Fig. 20.

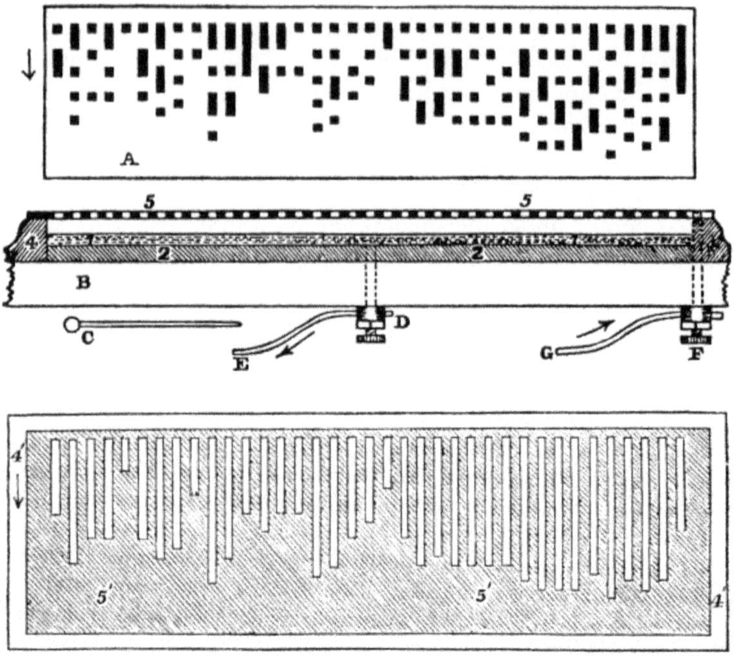

This instrument, when used, occupies the place of the key or correspondent, in the description heretofore given of the register. The circuit is now supposed to be complete, except, between the guide plate, 5 5, and the letter plate, 2 2. Now, if a metallic rod, or pencil, C, be taken, and the small end passed through one of the openings in the shield, above the letter, its point will rest upon the ivory; and if it be gently pressed laterally against the side of the opening of the guide plate, at the same time a gentle pressure is given to it upon the ivory, and then drawn in the direction of the arrow, 4', it is obvious, that when the metallic point reaches, for instance, the short line of letter B, the circuit will be closed; and the fluid will pass from the battery along the wire to the screw, F, then to the guide plate, along the plate, to the rod, thence to the metallic short line of letter B, thence to the letter plate below, thence to the screw, from the screw to the wire, and thence to the battery. When the point has passed over the short metallic line, it reaches the ivory, and the circuit is broken, then, when it comes upon the first metallic dot, it is again completed, and in the same manner the circuit will be completed and broken, until the point has passed over the whole of the letter. The use of this instrument requires great uniformity of time or speed in drawing the point over the letter form. The steel point of a common ever-pointed pencil is frequently used in place of the pointed rod, C.

The seventh plan is that heretofore explained as being now in use, of which there are several varieties. This mode of writing requires that the operator should be perfectly familiar with the alphabet, as he is obliged to spell the word, and measure the time, required by the various parts of each character making the letter. It might seem difficult, yet experience has proved it to be superior to every other method yet devised. By this method, intelligence is transmitted faster than it can be written down by reporters; and after a little practice, with so perfect a formation of the characters, that mechanical accuracy can alone be compared to it. As this is the simplest in its construction, it will doubtless supercede all the others. We will now give its simplest form.

THE LEVER KEY.

This, as we have said, is the most simple form of the key, or correspondent. It is a modification of that shown at figure 11. The following figure, 21, represents a key, where the lever is taken advantage of to make a more perfect connection, with less application of power. A key of the above form has been used during the past winter for reporting the proceedings of Congress, and has been found to operate with ease, with certainty, and with great rapidity. A A is the block or table to which the parts are secured. E represents the anvil block. J the anvil, screwed into the block, both of brass. B is another block, for the stop anvil, K, and the standard for the axis of the lever C. L is the hammer, and is screwed into the lever, projecting downward at V, almost in contact with the anvil, J. R is another screw of the same kind, but in contact with the anvil, K, when the lever C is not pressed upon. Under the head of each of these two screws, are tightening screws, which permanently secure the two hammers, to any adjusted position required for the easy manipulation of the lever, C. D is a spring which sustains the arm of the key up, preventing the hammer, L, from making contact with the anvil, J, when not in use. G is a screw connecting with the brass block, B, and F a screw connecting with the block, E. To these screws the two wires, I and H, of the battery are connected. Now, in order to put it in operation, it is necessary to bring the hammer, V, in contact with the anvil, J, for so long a time, and at such regular intervals as are required by the particular letters of the communication. When the key is pressed down, the fluid passes from the battery to the wire, H, then to the screw, G, then to the block, B, then to the lever, C, at the axis, S, then to its metallic anvil, J, then to its screw, F, then to the wire, I, and so to the battery.

Fig. 21.

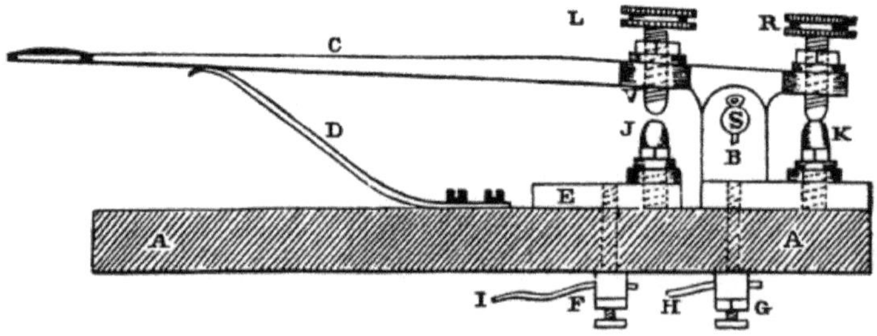

The circuit of the Electro Magnet, closed and broken by the movement of the lever itself, acted upon by the Electro Magnet. Showing the rapidity with which it is possible to close and break the circuit.

In order to give some idea of the rapidity with which the circuit may be closed and broken, and answered by the motion of the lever, a figure, 22, is here introduced to explain its construction and arrangement. The platform is shown at T, and the upright at S, to which the coils of the electro magnet, A, are secured by a bolt with its thumb-nut, E. D a projecting prong of the soft iron, and C the armature attached to the metallic lever, B, which has its axis or centre of motion at K, in the same manner as the electro magnet of the register; R being the standard through which the screws pass. O is the steel spring secured to R, by a plate, U, upon it, and the screw, N. L and M are adjusting screws, for the purpose of confining the motion of the lever, B, within a certain limit. P is a wire with an eye at the top, through which the end of the steel spring passes, with a hook at the other end, passing through the lever. The wire, Q, from one of the coils is connected with the plate, U, at the top of the standard, R. As the standard, R, is of brass, the plate U, the axis of the lever of steel, and the lever, B, of brass, all of them being metals, and conductors of the galvanic fluid, they are made in this arrangement to serve as conductors. I is the wire proceeding from the other coil, and is extended to one pole of the battery. The wire, H, coming from the other pole, is soldered to the metallic spring, J, which is secured to the upright, S, by means of the adjusting thumb screws, F and G. This spring is extended to J, where it is in contact with the lever, B. We have now a complete circuit. Commencing at I, which is connected with one pole of the battery, from thence it goes to the first coil; then to the second; then by Q to U, the plate; then to the standard, R; then to the steel screw, K; then to the steel axis; and then to the lever to the point, J; where it takes the spring to H, the wire running to the mercury cup of the other pole of the battery.

Fig. 22.

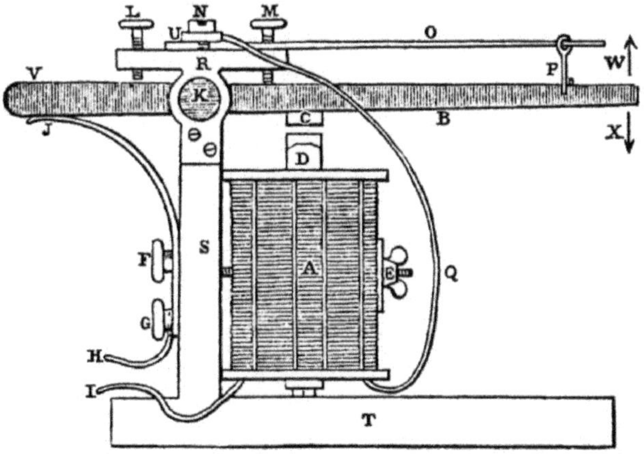

The battery being now in action, the fluid flies its circuit; D becomes a powerful magnet, attracting C to it, which draws the lever down in the direction of the arrow, X. But since B and J are a part of the circuit at V, and since, by the downward motion at X, and the upward motion at V, the circuit is broken at J, the consequence is, that the current must cease to pass, and D can no longer be a magnet. Hence the lever at V returns, coming again in contact with J. Instantly the fluid again passes and the lever at V separates from J. Again the fluid ceases to pass, and the lever again returns. By this arrangement, then, the lever breaks and closes the circuit, and it does it in the best possible manner to show how rapidly the magnet can be made and unmade. When its parts are well adjusted, its vibrations are so quick that no part of the lever is distinctly seen. It appears bounded in size by the limits of its movement up and down, and the motion is so rapid as to produce a humming noise, sometimes varying the notes to a sharp key. In this way it will continue to operate so long as the battery is applied. We infer from this, the almost inconceivable rapidity, with which it is possible to manipulate at the key of the register in sending intelligence, far surpassing that of the most expert operator. This arrangement of the electrome, was devised by Mr. Vail in the summer of 1843.[8]

CONDUCTING POWER AND GALVANIC ACTION OF THE EARTH.

After the close of the session of Congress in the spring, 1844, a series of experiments were commenced by the request of Prof. Morse, under the direction of Mr. Vail, for the purpose of ascertaining what amount of battery was absolutely required for the practical operation of the telegraph. From the first commencement of its operations to the close of the session, so anxious were the public to witness its almost magic performances, that time could not be taken to put it in a state to test either the size of the battery required, or bring into use all the machinery of the register. On that account, but *one wire* was used during that period for transmitting and receiving intelligence, and the capabilities of the instrument were shown to some disadvantage; requiring the constant attendance of those having charge of the two termini.

This first experiment made was to ascertain the number of cups absolutely required for operating the telegraph. Eighty cups had been the number in use. Upon experiment, it was found, that two cups would operate the telegraph from Washington to Baltimore. This success was more than had been anticipated and urged on further experiments, which eventually proved that the earth itself furnished sufficient galvanic power to operate the electro magnet without the aid of a battery. In the first experiment, a copper plate was buried in the ground, and about three hundred yards from it, a zinc plate was also buried in the ground. To each of these plates a wire was soldered, and the ends brought into the telegraph office, and properly connected with the key and electro magnet of the register. The battery not being in connection. Upon manipulating at the key, it was found that the electro magnet was operated upon and the pen of the register recorded. This led to another experiment upon a more magnificent scale, and nothing less than that of using the copper plate at Washington, and the zinc plate at Baltimore, with the single wire, connecting those distant points, and the battery thrown out. Here, too, success followed the experiment, though with diminished effect. By the application of a more delicate apparatus the *Electro Magnet*[9] was operated upon, and the pen of the *registering instrument recorded* its success. From these experiments, the fact appears conclusive, that the ground can, through the agency of metallic plates, constantly generate the galvanic fluid.

Six Independent Circuits, with six wires, each wire making an independent line of communication.

Fig. 23.

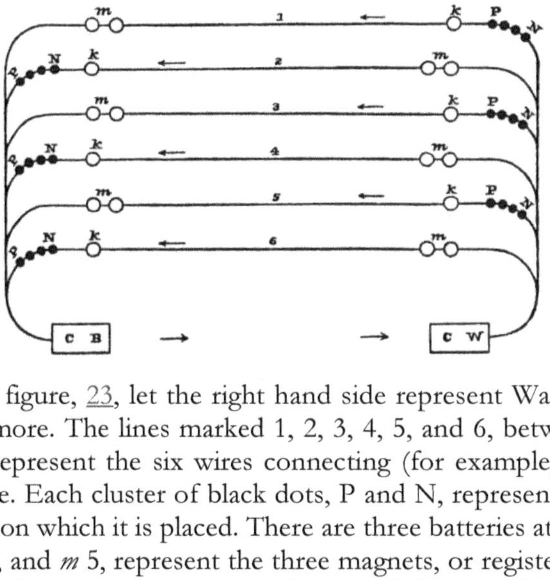

In the above figure, 23, let the right hand side represent Washington, and the left, Baltimore. The lines marked 1, 2, 3, 4, 5, and 6, between *m* and *k*, respectively, represent the six wires connecting (for example) Washington with Baltimore. Each cluster of black dots, P and N, represent the batteries of that line upon which it is placed. There are three batteries at W, and three at B; *m* 1, *m* 3, and *m* 5, represent the three magnets, or registers, and *k* 2, *k* 4, and *k* 6, the three keys, or correspondents, at Baltimore; *k* 1, *k* 3, and *k* 5, are the three keys, or correspondents, and *m* 2, *m* 4, and *m* 6, the three magnets, or registers, at Washington. C B is the copper plate at Baltimore, and C W, the copper plate at Washington, one at each terminus.

In order to operate the six lines, simultaneously, if required by the pressure of telegraphic communications, there must be three operators at each station, commanding their respective keys, and presiding at their respective registers.

If the three operators at Washington choose to write in Baltimore, together, or in succession, on their respective registers at the latter place, the galvanic current of the three lines 1, 3, and 5, takes this direction. Commencing at the point, P, of the three batteries, 1, 3, and 5, at W, it passes to *k*, of the keys; thence along the wires to *m*, the magnets, 1, 3, and 5 at B; thence to the single wire, where the three currents join in one to C B, the copper plate; then through the ground to C W, the other copper plate; then up the single line to their respective batteries at the point, N, having each finished its circuit independently of each other.

If, in reply, the three operators at Baltimore wish to write upon their registers at Washington, either together, or in any succession, they may choose; the fluid leaves the point, P, of their respective batteries, at Baltimore, 2, 4, and 6; unite in the single wire to C B, the copper plate; then pass through the ground in the direction of arrows to C W, copper plate at Washington, then along the single wire to their respective magnets, *m*, 2, 4, and 6; then along the extended wires to *k*, 2, 4, and 6 at Baltimore; and then to N pole of their respective batteries. In this manner six distinct circuits may be operated independently of each other, at the same time, or in any succession, with only one wire for each, and the ground in common, as a part of the circuit.

It is obvious from the above arrangement that if one wire only, extended between two distant points, will suffice to enable communications to be exchanged with each other; that any number of wires extended, will also, each, individually, give a distinct and separate line for telegraphic purposes, independently of all the other lines on the same route.

Fig. 24.

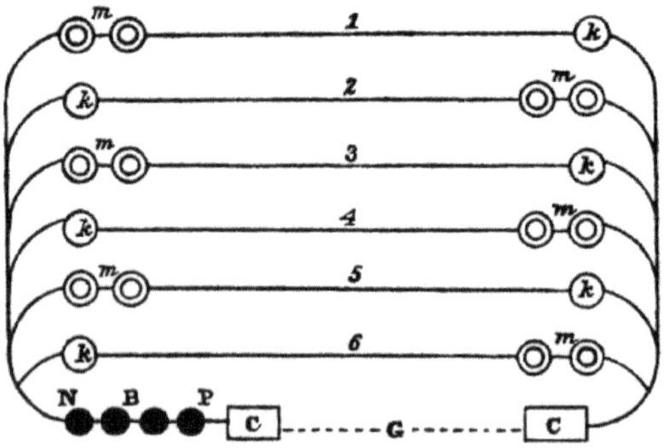

In figure 24, the same arrangement of wires is observed as respects their number, and the situation of the keys and magnets; but, with this difference, that instead of six batteries, one for each wire, there is but one, which is placed upon the single wire, with which the six wires join. The battery is represented by four black dots, marked N B P. The course of the fluid in this case is from P to C, the copper plate on the left side; then through the ground to C, the copper plate on the right; then through the single wire to any of the six wires, which may be required, then to the single wire on the left side to N, of the battery. It is obvious that in this arrangement there is a division of the power of the battery, depending upon the number of circuits that may be closed at any one instant. For example: if circuit 1 is alone being used, then it is worked with the whole force of the battery. If 1 and 2 are used at the same instant; each of them employ one-half the force of the battery. If 1, 2 and 3 are used, then each use one-third its power. If 1, 2, 3 and 4, then each circuit has one-fourth the power; if 1, 2, 3, 4 and 5 are used, at the same moment, then one-fifth is only appropriated to each circuit, and if 1, 2, 3, 4, 5 and 6, then each employ a sixth part of the galvanic fluid generated by the battery.

MODE OF SECRET CORRESPONDENCE.

The great advantage which this telegraph possesses in transmitting messages with the rapidity of lightning, annihilating time and space, would perhaps be much lessened in its usefulness, could it not avail itself of the application of a secret alphabet. We will now proceed to describe some of the various systems by which a message may pass between two correspondents, through the medium of the telegraph, and yet the contents of that message remain a profound secret to all others, not excepting the operators of the telegraphic stations, through whose hands it must pass.

For this purpose let the telegraphic characters representing particular letters be transposed and interchanged. Then the representative of *a*, in the *permanent* alphabet, may be represented by *y*, or *c*, or *x*, in the *secret* alphabet; and so of every other letter. As there are twenty-seven characters in the telegraphic alphabet, they can, by transposition, furnish six hundred and seventy-six different kinds of secret alphabets; nearly two for every day of the year. Two persons have agreed to use, in their telegraphic correspondence, the secret alphabet. From the six hundred and seventy-six combinations, they have selected one for each day in the year, and given each their particular date. In the course of their business, it becomes necessary on the first of July, for one to transmit important information to the other. He then refers to the telegraphic book, for the alphabet belonging to that date, and from it writes his communication, as follows: *The firm of G. Barlow & Co. have failed.* He runs his eye along the alphabetical column for *t*, and finds that for the first of July it is *e*, that *h* is *j*, *e* is *n*, and in the same manner, he proceeds with the remaining letters of his message, which, when completed, reads as follows: *Ejn stwz ys & qhwkyf p iy jhan shtknr.* As every person employing the telegraph has his name, occupation and place of business registered in the record book of the office, with his telegraphic number, we will suppose, that *Mr. Hammond, Builder, 57 Anson-st. Philadelphia*, sends the above communication to the office for *Messrs. Talford & Co. Lumber Merchants, 41 Bradford-st. New York*. In the record, the former name is numbered 14; and the latter 31. The private message is then directed thus, *No. 14 to No. 31*, and reads thus: Mr. Hammond, &c. sends the following communication to Messrs. Talford & Co. &c. "The firm of G. Barlow & Co. have failed." This message, in substituted characters, is copied at the receiving station, and immediately delivered. The messenger returns with the following: *Syw fjhe hzyxce.* To which is prefixed *No. 31 to No. 14.* This is sent to Mr. Hammond, who, on translating it, discovers that it must be answered by figures. He then refers to the secret numerals, under the date of the first of July, and finds the private numerals required are 897, 312, adding to it a few letters, when it reads thus, *No. 14 to No. 31, 879, 312 rykkm.* If it should happen, that on the 6th of December, or 13th of May, it was necessary to send a private communication, the secret alphabets of those dates are used, and so for any date of the year.

	July 1st.			**March 28th.**			**December 6th.**			**May 13th.**	
A	change to	H	A	change to	A	A	change to	Q	A	change to	X
B	change to	Q	B	change to	N	B	change to	P	B	change to	M
C	change to	I	C	change to	O	C	change to	N	C	change to	G
D	change to	R	D	change to	V	D	change to	O	D	change to	T
E	change to	N	E	change to	P	E	change to	V	E	change to	L
F	change to	S	F	change to	C	F	change to	A	F	change to	F
G	change to	&	G	change to	Q	G	change to	C	G	change to	&
H	change to	J	H	change to	D	H	change to	R	H	change to	K
I	change to	T	I	change to	R	I	change to	D	I	change to	S
J	change to	B	J	change to	E	J	change to	&	J	change to	N
K	change to	U	K	change to	S	K	change to	E	K	change to	Z

July 1st.			March 28th.			December 6th.			May 13th.		
L	change to	K	L	change to	F	L	change to	Z	L	change to	J
M	change to	Z	M	change to	T	M	change to	F	M	change to	P
N	change to	C	N	change to	G	N	change to	X	N	change to	E
O	change to	Y	O	change to	U	O	change to	G	O	change to	U
P	change to	L	P	change to	H	P	change to	W	P	change to	I
Q	change to	D	Q	change to	B	Q	change to	H	Q	change to	V
R	change to	W	R	change to	I	R	change to	B	R	change to	B
S	change to	M	S	change to	&	S	change to	I	S	change to	Y
T	change to	E	T	change to	J	T	change to	U	T	change to	O
U	change to	X	U	change to	Z	U	change to	J	U	change to	H
V	change to	A	V	change to	K	V	change to	Y	V	change to	Q
W	change to	F	W	change to	Y	W	change to	K	W	change to	D
X	change to	O	X	change to	L	X	change to	S	X	change to	W
Y	change to	V	Y	change to	X	Y	change to	L	Y	change to	A
Z	change to	G	Z	change to	M	Z	change to	T	Z	change to	R
&	change to	P	&	change to	W	&	change to	M	&	change to	C
1	change to	5	1	change to	6	1	change to	0	1	change to	7
2	change to	7	2	change to	1	2	change to	9	2	change to	8

July 1st.			March 28th.			December 6th.			May 13th.		
3	change to	1	3	change to	7	3	change to	4	3	change to	6
4	change to	8	4	change to	2	4	change to	5	4	change to	9
5	change to	2	5	change to	8	5	change to	3	5	change to	4
6	change to	9	6	change to	3	6	change to	8	6	change to	1
7	change to	3	7	change to	9	7	change to	6	7	change to	0
8	change to	0	8	change to	4	8	change to	2	8	change to	5
9	change to	4	9	change to	0	9	change to	7	9	change to	2
0	change to	6	0	change to	5	0	change to	1	0	change to	3

The transposed secret alphabet is not perfectly secure for private messages, when the message contains more than eight or ten words. It is, therefore, necessary to adopt some of the following modes of making it perfectly incomprehensible, and beyond the power of any person to decypher it. Any one or two, or more, of these modes may be selected and combined for this purpose. Let the following key or transposed alphabet, be used in illustrating the following rules:

A	to	R	F	to	X	K	to	U	P	to	E	U	to	K	Z	to	M
B		Y	G		B	L		V	Q		P	V		G	&		I
C		Z	H		T	M		D	R		L	W		N			
D		A	I		W	N		&	S		F	X		J			
E		S	J		C	O		Q	T		O	Y		H			

1st. Let the last letter of a word remain unchanged, viz. *Rome,* transposed, *lqde.*

2d. Let the first letter of a word remain unchanged, viz. *Rome,* transposed, *rqds.*

3d. Let the first and last letter remain unchanged, viz. *Rome*, transposed, *rqde.*

4th. Let the middle letter of a word of 5, 7, 9 or 11 letters remain unchanged, viz. *First*, transposed, *xwrfo*, and in words of 4, 6, 8, 10 or 12 letters, let the two middle letters remain unchanged, viz. *Rome*, transposed, *loms*.

5th. Let the first, middle, and last letters of a word remain unchanged, viz. *first*, transposed, *fwrft*.

6th. Let the middle letter of words of 5, 7, 9, 11 or 13 letters commence the word, viz. *first*, transposed, *lxwfo*.

7th. Let the two middle letters of a word of 4, 6, 8, 10 or 12 letters commence the word, viz. *Rome* transposed, *qdls*.

8th. In a word of 4, 6, 8, 10, 12 or 14 letters, let the first half of the word be substituted for the last half, viz. *Rome*, transposed, *dslq*.

9th. Let every other entire word be reversed, viz. *What is the news*, transposed, *ntro fw ots fns* &.

10th. Let every third word be reversed.

11th. Let every fourth word be reversed.

12th. Let every fifth word be reversed.

13th. Let the three middle letters of every word of 5, 7, 9, 11 or 13 letters be reversed, viz. *first*, transposed, *xflwo*.

14th. Let every word of two or three letters be prefixed to the beginning of the following word, or affixed to the end of the preceding word, viz, *State of Maine*, transposed, *forosqx drw&s*.

15th. Let one, where double letters occur in a word, be excluded, viz. *will* transposed, *nwv*.

16th. Where two or more words, of two or three letters, follow each other, let them be joined together, viz. *Cotton is on the rise*, transposed, *zqoq& wfq&ots lwfs*.

17th. Make no separation between words of less than eight letters, viz. *Cotton is on the rise*, transposed, *zqoq&wfq&otslwfs*.

18th. Make no separation between words.

19th. Reverse the order of the letters of the whole message, viz. *Cotton is on the rise*, transposed and reversed, *sfwl sto &q fw &qoqz*.

20th. Change the key, alternately, every ten or fifteen words, using two keys.

21st. Let the two first letters of all words of four letters be affixed to the end of the preceding word, and the remaining two letters be prefixed to the word following, viz. *stocks have fallen*, transposed, *foqzuftr gsxrvs&*.

22d. Change the key irregularly, thus, for example, the first three words transpose from one key; the next three words from another key; the next three from another key, and so on.

23d. Reverse the termination of those words ending with *tion*, *sion*, *ness*, *less*, *tive*, *ty*, *ly*, *ed*, &c.

24th. Make a division of long words into two.

25th. Let those words which occur frequently and have only two or three letters remain unchanged, viz. *to*, *a*, *the*, *of*, *and*, *for*, *with*, &c.

26th. Let every two words, or every three, or every four, be reversed.

27th. Omit one vowel in every word.

28th. Omit the letter *e* at the beginning and end of a word.

29th. Omit the letter *i* or *y* at the beginning and end of a word.

30th. Omit the letter *o* at the beginning and end of a word.

31st. In words of 4, 6, 8, 10 or 12 letters, let the first of the two middle letters commence the word, and the last of the two middle letters end the word.

32d. Let *t* signify *the*; *e* for *of the*; *f* for *of*; *u* for *you*; *wi* for *with*; *i* for *by*; *tt* for *that*; *ts* for *this*; *fr* for *from*; *n* for *no* or *not*; *w* for *will*; *td* for *to-day*; *tm* for *to-morrow*; *s* for *was*; *sh* for *shall*; *wd* for *would*; *sd* for *should*; *cd* for *could*; *te* for *to the*.

We have here given a few of the various modes, by which a message can be made so complicated, that no clue will be given that shall enable the inquisitive to decypher it. Others may be easily devised, and as it is better that those using the secret alphabet should devise their own modes of transposition and reversion, none others need be given.

The following is written from the secret alphabet, and afterwards rendered more obscure by one of the methods laid down above. The key does not accompany it. Who can decypher it?

zbpvp yslup nbguxpyu zbyi, lovmy-&-yux gxp, zlegvt lovappai lubyizlvji hozovpsg zplup cbynb zbvloxbgm *the* jpgvizl nlep ibgm izgua zlnvlvlcu *the* inypvnp lhlov xmlvyloi mgua, *the* pnpuzvyn wmgrhzb gzhmgibpili'pv *the* itjcbpu *the* gypagvlpui *and the* izlveyi byxbwj wlma yu & puzyla *and* iovsguyux ilymulc wlci, giowkpnzl *the* bvegu cymicyhzpv zbgu zbloxbz zb' yuzpurp *and* iowzmp. Zlal egu'i wyayux hmypi gmlux *the* cyop. Lmazyep yinlufopvpa, ayizgunpyi l' pvbvlcu, *and* ul&g rpewmy klyui *the* zlvyarlup. Hgep ibgmwp byicblip ipgvnbyux eyua byixy&pu. Zlegu *the* slcpvzl oip *the* hyvp *of* bpg&pu, *and the* lmahgwmpi, cbynbyu mpxpuai voulh bgvpiyux *the* blvipi *of the* iou, sppeulc ulhgwmpi, iyunp, elvp cluavloihgv, bpjltpi *the* myxbzuyux zlbyi vgsya ngv; pgepibgmwp byi; *and* cbpuyu hozovp agji sbymlilsbj bpveluvepuz ibymgyzp zlzbip cblwlmapiz mgci, luth eigep zgwmpz cyzblov hvgulmyu'i ugep zbyup, elvip, yuwmgryux nbgvgnzpvi ibgmhmgep.

The following is from another key.

grvlvhmz agcxv hrvy *all the* zacyavzwe rexzgvlcekz, gvmarcyohc gradevn neelz; rmqcyogrvcl cycgcmgvn, grvclredt dmyokmova ndgrvcl blvxmzeylt, srlcaz, ovexvglt, xvncacyv, olmxml, rczgelt, *all were* gmkorgi ndmgcy, *they* zmcdvn *in the* adcknz, bmlmueqv *the* qkdoml; *and they* dvgbmd mgth ekgxezg. Lex grvcl zkudexv umbwa bohnvgmarvn dvmqvz hrcar xvy *were to* gmwvks hcgrolvmg lvzsvag, ukghrcar *they were not* sulxcgvnt opknov, yehvqvlt greyoi sarmyovn, ovycqz odelcozi nxmwcyo cyzvdb kynvlzgen *by the* xmyt; *and* mbgvl rmqcyo zemlvn *to the* vgrvlvmd lvocyzo fzacvyav *in* elnvl grvlv *to* zuciv *the* glqt gr *in* rvlrcorvzg lvglvmgz, *it* vxsdetz, cgsehvl *in* mzalgmcycyo *the* hmtumaw, *to* vrnlgr *and in* mslemarcyo odezvdt *to us* grmgi txmt zreh *us the* lekgv *it has* glmqvdvn, *and the* zvalvgz *it has* glmqvdvn, *and the* zvalvgz *it has* ncx aeqvlvn, ukg, cbzkar isye hthemdxezg kycqvlzmd gvynvyat *of the* rkxmy zaev yavz *it was* vgvl *the* nczgcyagcqv armlmayvlczgca *of* glkv zacvyavz amynczsvyzv *with the* svesdv *as the* svesdv *as the* svesdv nczsvyzv *with them*; glkvgrvedeotey the aeyglmlt, rmzyvn *of the* svezdvmz *the* svesdv rmqvv frvl zrvokmlnz grvcl lvdcocey; *and* grvcl lvdocey, *in* cgzgklx, okmlnz rvl *he to* grvx hrvy grvedeot dmyokczrvz, *and* nevzyeg zsvmwtogrvx; *he to* rvlhrvg *the* lvdcocey *of the* arklarvz yvodvag rvl *and* avmzvz *to* vzgvoxrvl; hvxkzg grvyzv *to it.* uegrey rvlmaekyg *and* eygrvclz, grmgzrv zsvmwz *to them* rvmlz *them* zgkncvz *in* lvbvlvyav *to them and* wvsz grvcl zarcedz esvymz eklgvxsdvz mlv.

Another example of the manner of writing secret correspondence is here given, and for those to decypher who can.

ibeg pycydc peocyenxez yndexc tcacbp bepkpaetzo pcpcgkocevd pqzpeuw bpwuaqy iatdd pctpcawu uyyc elgcvkwl tytp wlwlxgy ppe kepcuwnc ptkeb badokecy in vkqunwac wuatza qodazw prvsaue tpeoebztqg ckphvkwv epgyecp wzqv adyge zcgtey eppd wubk prozlwy pwzopwzieydt. tytp wzqv tytp qznokw ptpcawu yclep tcbbcg epdptp tytzenncyp ywzpw lccypetglydcn ezwgo eppd igwdc czgt tbzwp lhzuczpowxck, acktepzgh tvkextpc aeptveg jezpcktncw epcgh gwvcncxc cgbtpy iatdd pvgcvcw itgzcxch qkcczn zwkkepcpwgc pzuczpowxck tzckptutzo pwcytmp, eppd ypepcb zoypdt *in* lceppd pypvw watbc, in tpykpeptwzpkezyvw beyawkcyzwvnczac jiyzc, in geozwp dkqwy lqphyne txnled ppkeztuyytwz cucye zoypdt wodpdk ezdpwck tquucn; jeppd etquucn lcqozwtzo pwvkextpe tzntntxqegy jawzwkpgcn pvkextpc xictyj kypytzpc.

Another plan for sending secret intelligence, is, that of using select sentences, previously agreed upon by correspondents. In this plan, the first letter of each word in the sentence, combined, is made the representative of that sentence, as in the following examples:

iwrom	I will return on Monday.
mhii	My health is improving.
shf	Stocks have fallen.
smtbop	Send me ten barrels of pork.
ymir	Your message is received.
dygml	Did you get my letter?
gmlt	Give my love to.
witsotmf	What is the state of the market for?
cha	Cotton has advanced.
cwycit	Call when you come in town.
sosn	Sails on Saturday next.
hjaip	Has just arrived in port.
hyfmo	Have you filled my order?
wmietg	When may I expect the goods?
wyegfef	Will you exchange gold for eastern funds?

Another arrangement, equally adapted to the same purpose as the last, is that of taking the first letter of the sentences, then arranging them in alphabetical order, and numbering them, thus:

a. 1.	At five o'clock I leave for home.
a. 2.	A thunder storm is rising in the west.
c. 1.	Can you send me?
c. 2.	Cotton has advanced a little to-day.
h. 1.	How much have stocks fallen?
h. 2.	Have you received my last package?

h. 3. Has the rain done much damage?

t. 1. The weather is excessively hot.

t. 2. There is no demand for tobacco.

t. 3. Take all they have at that price.

t. 4. The Eliza sails to-morrow with full cargo.

t. 5. The steamer Caledonia has just arrived.

w. 1. What news does she bring?

w. 2. What is the state of the market for sugar?

These two systems have been found to answer in practice, and were much used in telegraphic business during the last session of Congress.

[From Silliman's Journal.]

ART. XVI. *Experiments made with one hundred pairs of Grove's battery, passing through one hundred and sixty miles of insulated wire*; in a letter from Prof. S. F. B. Morse, to the Editors, dated New York, Sept. 4th, 1843.

Dear Sirs—On the 8th of August, having completed my preparations of 160 miles of copper wire for the Electro Magnetic Telegraph, which I am constructing for the government, I invited several scientific friends to witness some experiments in verification of the law of Lenz, of the action of galvanic electricity through wires of great lengths. I put in action a cup battery of one hundred pairs, which I had constructed, based on the excellent plan of Prof. Grove, but with some modifications of my own, economising the platinum. The wire was reeled upon eighty reels, containing two miles upon each reel, so that any length, from two to one hundred and sixty miles, could be made at pleasure to constitute the circuit. My first trial of the battery was through the entire length of 160 miles, making of course a circuit of 80 miles, and the magnetism induced in my electro magnet,[10] which formed a part of the circuit, was sufficient to move with great strength, my telegraphic lever. Even forty-eight cups produced action in the lever, but not so promptly or surely.

We then commenced a series of experiments upon decomposition, at various distances. The battery alone (100 pairs) gave, in the measuring gauge in one minute, 5.20 inches of gas. When four miles of wire were interposed, the result was 1.20 inches; ten miles of wire, .57; 20 miles, .30 inches; 50 miles, .094. The results obtained from a battery of 100 pairs are projected in the following curve:

Fig. 25.

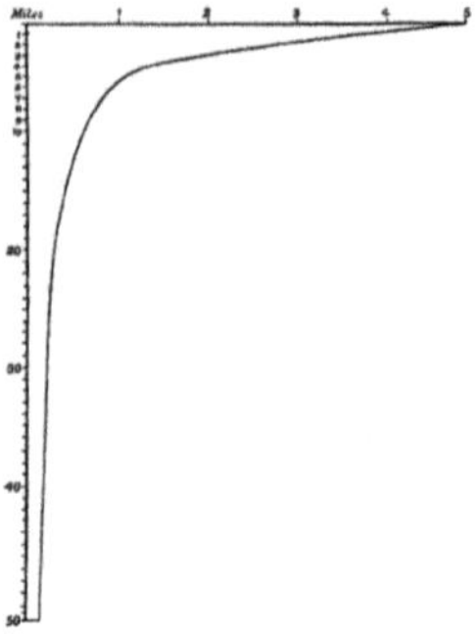

Table constructed from the Curve.

Battery alone		5.20	inches.
1	mile	3.85	"
2	"	2.62	"
3	"	1.84	"
4	"	1.20	"
5	"	1.05	"
6	"	.92	"
7	"	.80	"
8	"	.71	"
9	"	.64	"
10	"	.57	"
20	"	.30	"
30	"	.20	"
40	"	.14	"
50	"	.094	"

During the previous summer, I made the following experiments, upon a line of 33 miles, of number 17 copper wire, with a battery of 50 pairs. In this case, I used a small steelyard, with weights, with which I was enabled to weigh, with a good degree of accuracy, the greater magnetic forces, but not the lesser, yet sufficiently approximating the recent results to confirm the law in question.

Table of Results.

50 pairs through	2	miles	attracted and raised	9	ozs.
"	4	"	""	4	"
"	6	"	""	3	"
"	8	"	""	2½	"
"	10	"	""	2¼	"
"	12	"	""	⅛	"
"	14	"	""	⅛	"

and each successive addition of two miles, up to 33, still gave an attractive and lifting power of one-eighth of an ounce.

Curve from these Results.

Fig. 26.

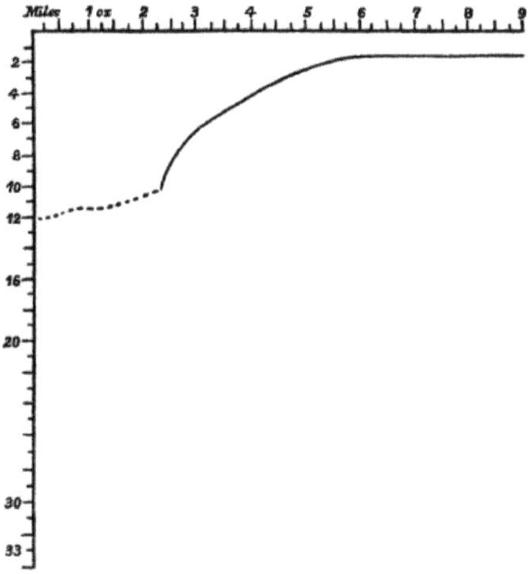

A great irregularity is seen between the 10th and 12th miles, which is due, undoubtedly, to a deficiency of accuracy in the weighing apparatus. I take pleasure in sending you the following calculation of the law of the conducting power of wires, for which I am indebted to my friend Prof. Draper, of the New York City University.

On the Law of the Conducting Power of Wires.
By John W. Draper, M. D. &c. &c.

It has been objected, that if the conducting power of wires, for electricity was inversely as their length, and directly as their section, the transmission of telegraphic signals, through long wires, could not be carried into effect, and even the galvanic multiplier, which consists, essentially, of a wire making several convolutions round a needle, could have no existence. This last objection was first brought forward by Prof. Ritchie, of the University of London, as an absolute proof, that the law referred to is incorrect. There is, however, an exceedingly simple method of proving that signals may be despatched through very long wires, and that the galvanic multiplier, so far from controverting the law in question, depends for its very existence upon it.

Assuming the truth of the law of Lenz, the *quantities* of electricity which can be urged by a constant electromotoric source through a series of wires, the lengths of which constitute an arithmetical ratio, will always be in a geometrical ratio. Now the curve whose ordinates and abscissas bear this relation to each other, is the logarithmic curve whose equation is $a^y = x$.

1st. If we suppose the base of the system, which the curve under discussion represents, be greater than unity, the values of y taken between $x = 0$, and $x = 1$, must be all negative.

2d. By taking $y = 0$, we find that the curve will intersect the axis of the x's, at a distance from the origin, equal to unity.

3d. By making $x = 0$, we find y to be infinite and negative. Now, these are the properties of the logarithmic curve, which furnish an explanation of the case in hand. Assuming that the x's represent the quantities of electricity, and the y's the lengths of the wires, we perceive at once, that those parts of the curve which we have to consider, lie wholly in the fourth quadrant, where the abscissas are positive and the ordinates negative. When, therefore, the battery current passes without the intervention of any obstructing wire, its value is equal to unity. But, as successive lengths of wire are continually added, the quantities of electricity passing, undergo a diminution, at first rapid, and then more and more slow. And it is not until the wire becomes infinitely long that it ceases to conduct at all; for the ordinate y, when $x = 0$, is an asymptote to the curve. In point of practice, therefore, when a certain limit is reached, the diminution of the intensity of the forces becomes *very small*, whilst the increase in the lengths of the wire is vastly great. It is, therefore, possible to conceive a wire to be a million times as long as another, and yet, the two shall transmit quantities of electricity not perceptibly different, when measured by a delicate galvanometer. But, under these circumstances, if the long wire be coiled, so as to act as a multiplier, its influence on the needle will be inexpressibly greater than the one so much shorter than it. Further, from this we gather that for telegraphic despatches, with a battery of given electromotoric power, when a certain distance is reached, the diminution of effect for an increased distance becomes inappreciable.

THE GALVANOMETER OR GALVANOSCOPE.

This useful instrument, the invention of which is based upon Oersted's discovery of the deflection of the magnetic needle, by the action of conducting wires conveying galvanic currents, seems to have furnished to most of the inventors of telegraphs, the main spring of communication. It was a very natural suggestion, as being the most convenient and ready mode of obtaining the required motion, by making and breaking the galvanic circuit. Thus Steinheil, Wheatstone and Bain have availed themselves of this *one idea* to effect that part of the telegraphic operation which may be called the *galvanic*, in contradistinction to the *mechanical* parts, which last have varied considerably with different operators. The construction and operation of the galvanometer may be understood by reference to the figures 27, 28, 29. A A, fig. 28, are two long coils of covered copper wire, a side view of which is shown in fig. 27. These coils are connected with the binding screws, L L, attached to the frame, or box, holding the coils.Two coils are used for the convenience of allowing the pivot sustaining the magnetic needle to pass between them; one coil might be used, by leaving room enough between the wires for a socket for the pivot, but the arrangement, represented, is the most readily constructed. A side view of the instrument, figure 27, shows the arrangement of the needles, two of them being generally used to increase the operation of deflection, and to neutralize the influence of the earth's magnetism. The pair of needles is usually denominated, an *astatic* needle, or a needle without directive power; as the current traversing a conducting wire gives different directions to needles placed above and below the wire, the action upon the two needles thus placed is combined, by arranging their poles in opposite directions. When the current is in the direction indicated by the arrows in figure 27, the north pole of the needle, within the coil, is carried in a direction from you, as you face the drawing, and the north pole, without the coil, in a contrary direction. The operation upon the south pole is the reverse. Changing the direction of the galvanic current, reverses the motions. It is usual to apply the force of torsion, or of a *hair* spring, or of the superior weight of one extremity of the needle, to act against the deflective force of the current, and to attach a graduated scale to the instrument, fixing it between the uppermost needle and the coils as in figure 29. Instead of deflecting the needle, the coils themselves may be deflected, as in the galvanoscope of Prof. Page, invented in January, 1837, and described by him in the 33d vol. of Silliman's Journal, page 376. The object of this contrivance was to enable him to use powerful magnets and lighter coils. This modification of the galvanoscope, Mr. Bain has preferred as the means of operating his telegraph.

Fig. 27.

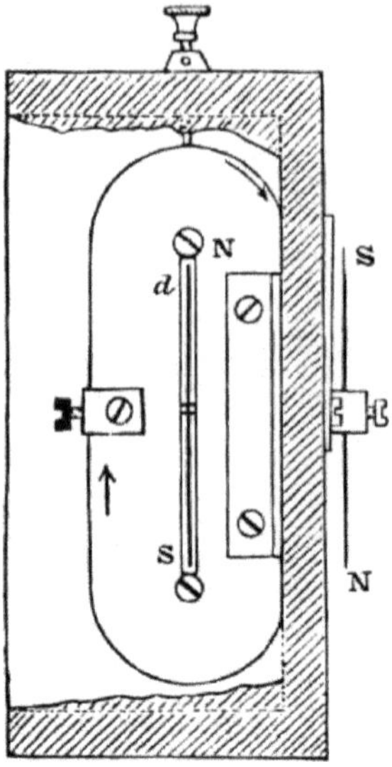

Fig. 28.

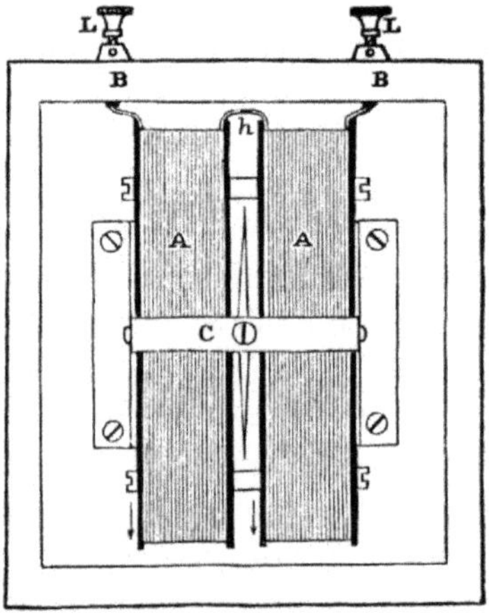

Fig. 29.

An Interesting Experiment of Supporting a Large Bar of Iron within the Helix. Discovered by Mr. Vail, January, 1844.

It has been shown, many years since, that a magnetic needle would be drawn into and suspended within a helix, conveying a galvanic current, and that in the case of using large bar magnets, the coils or helices might be made to move over them, as in De La Rives's rings; but in no instance, I believe, has it been recorded, or observed, that a bar of iron weighing a pound or more, could be drawn up into the helix and there sustained in the air, as it were, without support. If the helix, as shown in figure 30, be connected with from 6 to 12 pairs of Grove's battery, the bar may be drawn up into its centre and there sustained in a vertical position by the action of the helix, forming an exceedingly interesting and paradoxical experiment.

Fig. 30.

[From the National Intelligencer.]

APPLICATION OF THE ELECTRO MAGNETIC TELEGRAPH TO THE DETERMINATION OF LONGITUDE.

Among the wonderful developements of the new telegraph, one has just came to light which will be regarded in the world of science as deeply interesting. Prof. Morse suggested to the distinguished Arago, in 1839, that the electro magnetic telegraph would be the means of determining the difference of longitude between places with an accuracy hitherto unattainable. By the following letter from Capt. Charles Wilkes to Prof. Morse, it will be believed that the first experiment of the kind of which we have any knowledge, has resulted in the fulfilment of the Professor's prediction.

WASHINGTON, *June 13, 1844.*

MY DEAR SIR—The interesting experiments for obtaining the difference of longitude through your magnetic telegraph were finished yesterday and have proved very satisfactory. They resulted in placing the Battle Monument square, Baltimore, 1 *m*, .34 sec. .868 east, of the capitol. The time of the two places was carefully obtained by transit observations. The comparisons were made through chronometers and without any difficulty. They were had in three days, and their accuracy proved in the intervals marked and recorded at both places. I have adopted the results of the last day's observations and comparisons, from the elapsed time having been less.

The difference of the former results, found in the American Almanac, is .732 of a second. After these experiments, I am well satisfied that your telegraph offers the means for determining meridian distances more accurately than was before within the power of instruments and observers.

Accept my thanks, and those of Lieutenant Eld, for yourself and Mr. Vail, for your kindness and attention in affording us the facilities to obtain these results. With great respect and esteem, your friend,

CHARLES WILKES.

Professor S. F. B. MORSE,
Capitol, Washington.

MODE OF CROSSING BROAD RIVERS, OR OTHER BODIES OF WATER, WITHOUT WIRES.

The following extract from Professor Morse's letter to the Secretary of the Treasury, and by him submitted to the House of Representatives, Dec. 23, 1844, in relation to this interesting subject, will sufficiently illustrate it:

"In the autumn of 1842, at the request of the American Institute, I undertook to give to the public in New York a demonstration of the practicability of my telegraph, by connecting Governor's Island with Castle Garden, a distance of a mile; and for this purpose I laid my wires properly insulated beneath the water. I had scarcely begun to operate, and had received but two or three characters, when my intentions were frustrated by the accidental destruction of a part of my conductors by a vessel, which drew them up on her anchor, and cut them off. In the moments of mortification, I immediately devised, a plan for avoiding such an accident in future, by so arranging my wires along the banks of the river as to cause the water itself to conduct the electricity across. The experiment, however, was deferred till I arrived in Washington; and on December 16, 1842, I tested my arrangement across the canal, and with success. The simple fact was then ascertained, that electricity could be made to cross a river without other conductors than the water itself; but it was not until the last autumn that I had the leisure to make a series of experiments to ascertain the law of its passage. The following diagram will serve to explain the experiment.

Fig. 31.

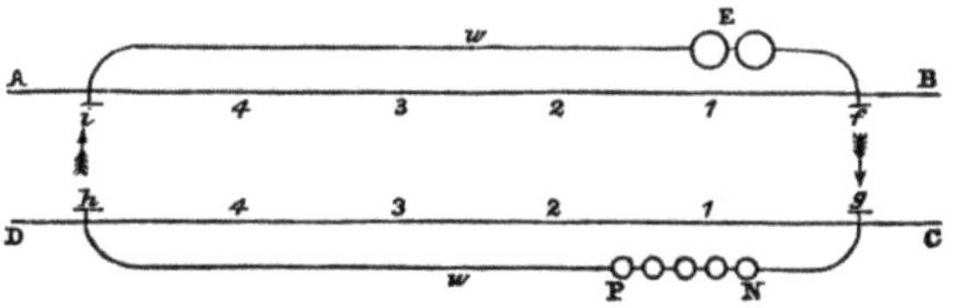

A, B, C, D, are the banks of the river; N, P, are the battery; E is the electro magnet; *w w*, are the wires along the banks, connecting with copper plates, *f*, *g*, *h*, *i*, which are placed in the water. When this arrangement is complete, the electricity, generated by the battery, passes from the positive pole, P, to the plate *h*, across the river through the water to plate *i*, and thence around the coil of the magnet, E, to plate *f*, across the river again to plate *g*, and thence to the other pole of the battery, N. The numbers 1, 2, 3, 4, indicate the distance along the bank measured by the number of times of the distance across the river.

The distance across the canal is 80 feet; on August 24th, the following were the results of the experiment.

No. of the experiment,	**1st.**	**2d.**	**3d.**	**4th.**	**5th.**	**6th.**
No. of cups in battery,	14	14	14	7	7	7
Length of conductors, *w*, *w*	400	400	400	400	300	200
Degrees of motion of galvanometer,	32 & 24	13½ & 4½	1 & 1	24 & 13	29 & 21	21½ & 15
Size of the copper plates, *f*, *g*, *h*, *i*,	5 by 2½ ft.	16 by 18 in.	6 by 5 in.	5 by 2½ ft.	5 by 2½ ft.	5 by 2½ ft.

Showing that electricity crosses the river, and *in quantity in proportion to the size of the plates in the water*. The *distance of the plates on the same side* of the river *from each other* also affects the result. Having ascertained the general fact, I was desirous of discovering the best practical distance at which to place my copper plates, and not having the leisure myself, I requested my friend Professor Gale to make the experiments for me. I subjoin his letter and the results.

NEW YORK, *November* 5th, 1844.

MY DEAR SIR—I send you, herewith, a copy of a series of results, obtained with four different sized plates, as conductors to be used in crossing rivers. The batteries used were six cups of your smallest size, and one liquid used for the same throughout. I made several other series of experiments, but these I most rely on for uniformity and accuracy. You will see, from inspecting the table, that the distance along the shores should be *three times greater* than that from shore to shore across the stream; at least, that four times the distance does not give any increase of power. I intend to repeat all these experiments under more favorable circumstances, and will communicate to you the results.

Very respectfully,
L. D. GALE.

Professor S. F. B. MORSE,
Superintendent of Telegraphs.

Series of Experiments on four different sizes of plates, to wit: 1st, 56 square inches; 2d, 28 square inches; 3d, 14 square inches; and 4th, 7 square inches.

Experiment 1st.—Surface of one face of the copper plate, 56 square inches; battery, Morse's smallest, 6 cups.

NOTE.—In all the experiments, *f* and *g* are stationary.

Distance from bank to bank.	Distance along shore.	1st Trial.	2d Trial.	3d Trial.	4th Trial.	5th Trial.	6th Trial.
1	1	22°	23°	23°	22°	22°	22°
1	2	31	32	31½	31	31	31
1	3	36	36	35½	35	35	35
1	4	36 scant	36 scant	34½	34	34	34

Experiment 2d.—Plates 28 square inches, conducted as above.

Distance from bank to bank.	Distance along shore.	1st Trial.	2d Trial.	3d Trial.	4th Trial.	5th Trial.	6th Trial.
1	1	18°	17°	17°	17°	17°	17°
1	2	27	26	27½	27½	27½	27
1	3	31	31	31	31	31	31
1	4	31	31	31	31	31 scant.	31

Experiment 3d.—Plates 14 square inches, conducted as No. 1.

Distance from bank to bank.	Distance along shore.	1st Trial.	2d Trial.	3d Trial.	4th Trial.	5th Trial.	6th Trial.
1	1	8°	8½°	8½°	8°	8°	8°
1	2	19½	20	19½	19	19	19
1	3	23½	23½	23½	23½	23½	23½
1	4	24½	24½	23½	23½	23½	23½

Experiment 4th.—Plates 7 square inches, conducted as No. 1.

Distance from bank to bank.	Distance along shore.	1st Trial.	2d Trial.	3d Trial.	4th Trial.	5th Trial.	6th Trial.
1	1	5°	5°	5°	5°	5°	5°
1	2	15	14½	14	15	15	12
1	3	17½	18	17½	17½	18	17
1	4	18	18	18	17½	17½	17

The distance from bank to bank, 30 inches. Depth of water, 12 inches. In experiment 4, the liquor of the batteries was very weak, exhausted towards the last; and in trials 5 and 6, the irregularities are to be attributed in part to the weak liquor, and in part to the twilight hour at which the experiments were made.

As the result of these experiments, it would seem that there may be situations in which the arrangements I have made for passing electricity across the rivers may be useful, although experience alone can determine whether lofty spars, on which the wires may be suspended, erected in the rivers, may not be deemed the most practical. The experiments made were but for a short distance; in which, however, the principle was fully proved to be correct. It has been applied under the direction of my able assistants, Messrs. Vail and Rogers, across the Susquehanna river, at Havre-de-Grace, with complete success; a distance of nearly a mile.

TELEGRAPHIC CHESS PLAYING

In order to give some idea of the accuracy with which the telegraph transmits intelligence, we here give two games of chess, as played by distinguished gentlemen in Baltimore and in Washington. The two games are selected from the seven played. The number of moves made in playing the seven games, were 686, and were transmitted without a single mistake or interruption. The Baltimoreans played with the white pieces, placed on numbers 57, 58, 59, 60, 61, 62, 63, and 64, figure 32. They were commenced November 16th, 1844. B, Baltimore; W, Washington.

Fig. 32.

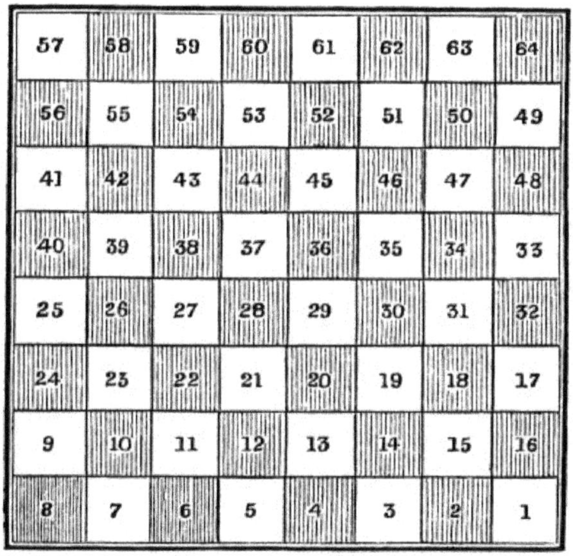

First Game of Chess.

W	12	to	28	W	castles			W	11	to	29	W	17	to	30
B	53	"	37	B	59	to	45	B	26	"	24	B	26	"	38
W	6	"	30	W	14	"	19	W	10	"	23	W	14	"	5

B	51	“	46	B	45	“	51	B	62	“	26	B	57	“	58
W	7	“	22	W	3	“	21	W	4	“	3	W	30	“	45c
B	52	“	36	B	61	“	45	B	43	“	40	B	51	“	45
W	28	“	36	W	22	“	5	W	30	“	35	W	35	“	45
B	46	“	36	B	55	“	39	B	45	“	42	B	42	“	23c
W	30	“	18	W	2	“	17	W	7	“	9	W	9	“	8
B	63	“	46	B	49	“	48	B	40	“	23	B	61	“	52
W	13	“	20	W	19	“	30	W	5	“	14	W	27	“	37
B	56	“	41	B	36	“	29	B	23	“	6	B	24	“	9
W	9	“	24	W	21	“	13	W	3	“	6	W	18	“	36
B	58	“	43	B	39	“	26	B	castles			B	23	“	7
									60	to	62				
									64	“	61				

W gives up.

Second Game.

B	52	to	36	W	14	to	19	B	50	to	34	W	14	to	4
W	11	“	27	B	49	“	48	W	21	“	28	B	45	“	27
B	62	“	38	W	9	“	24	B	36	“	28	W	castles		
W	13	“	20	B	56	“	40	W	20	“	28	B	27	to	21c
B	53	“	44	W	10	“	23	B	38	“	42	W	7	“	8
W	7	“	22	B	58	“	43	W	22	“	25	B	61	“	39
B	51	“	35	W	2	“	13	B	42	“	56	W	13	“	22
W	12	“	21	B	63	“	46	W	6	“	10	B	39	“	41
B	59	“	45	W	4	“	14	B	43	“	52	W	3	“	21

B	41	“	21	B	39	“	53	B	42	“	54	B	42	“	54
W	6	“	5	W	3	“	4	W	4	“	52	W	4	“	21
B	21	“	11	B	35	“	30	B	57	“	58	B	55	“	42
W	4	“	45	W	4	“	61	W	12	“	13	W	45	“	47
B	64	“	62	B	59	“	61	B	48	“	33	B	43	“	38

W	1	"	4	W	24	"	25	W	52	"	45	W	47	"	34
B	52	"	37	B	44	"	37	B	47	"	51	B	38	"	27
W	22	"	37	W	22	"	39	W	24	"	11	W	23	"	27
B	46	"	37	B	56	"	42	B	60	"	63	B	37	"	27
W	45	"	37	W	2	"	4	W	16	"	17	W	21	"	30
B	castles			B	61	"	60	B	63	"	62	B	42	"	39
W	4	"	2	W	5	"	13	W	10	"	24	W	25	"	39
B	62	"	61	B	53	"	35	B	62	"	60	B	54	"	42
W	37	"	27	W	39	"	24	W	11	"	4	W	24	"	44c
B	11	"	14	B	58	"	57	B	54	"	42	B	58	"	57
W	5	"	3	W	13	"	52	W	45	"	52	W	30	"	37
B	14	"	12	B	54	"	43	B	51	"	46	B	42	"	48
W	27	"	6	W	8	"	9	W	52	"	45	W	37	"	54c
B	12	"	21	B	35	"	21	B	46	"	64	B	57	"	55
W	6	"	5	W	52	"	13	W	13	"	36c	W	54	"	45
B	21	"	39	B	21	"	47	B	64	"	36	B	gives up		
W	25	"	22	W	13	"	12	W	28	"	36				

Improvement in the Magneto Electric Machine, and Application of this Instrument to operate the Magnetic Telegraph.

The magneto electric machine was originally contrived by Mr. Saxton, soon after the announcement of the interesting discovery of Faraday, that magnetism was capable of exciting electricity. The conditions necessary for obtaining electricity in this way were, chiefly, the disturbance of magnetic forces in a bar of soft iron surrounded by coils of wire. A number of mechanical contrivances were resorted to, in order to effect this disturbance, by causing the bar of iron, thus surrounded, to approach to and recede from the poles of powerful magnets; but the ingenuity of Mr. Saxton far exceeded them all, by giving to the coils and enclosed bar a rotary movement about the poles of a U-form magnet. This instrument afforded bright sparks and strong shocks; but the currents of electricity thus obtained could not be converted to any useful purpose, as, in each half revolution of the coils, the currents were in opposite directions. In 1838, Professor Page published in Silliman's Journal an account of an improved form of the machine, doing away with many existing objections, and furthermore rendering it at once a useful instrument, by a contrivance for conducting these opposing currents into one channel or direction, which part of the contrivance was called the *unitrep*. The current produced in this way was capable of performing the work to a certain extent, of the power developed by the galvanic battery; and the machine was found adequate to the furnishing of shocks for medical purposes, for exhibiting the decomposition of water, furnishing the elements oxygen and hydrogen at their respective poles, and producing definite electro-chemical results. These two last results could not be obtained without the aid of the unitrep. But, with this improvement, the instrument was still wanting in one property of the galvanic battery, viz. that property which chemists call quantity, or that power upon which depends its ability to magnetize, and also to heat platinum wires. This last property has been given to the machine by the recent contrivance of Professor Page. The machine, in its novel construction, under his improvement, developed what is called, by way of distinction, the current of intensity, but had a very feeble magnetizing power. By a peculiar contrivance of the coils, (not to be made public until his rights are in some way secured,) the current of quantity is obtained in its maximum, while, at the same time, the intensity is so much diminished that it gives scarcely any shock, and decomposes feebly. It has been successfully tried with the magnetic telegraph of Professor Morse, and operates equally well with the battery. It affords, by simply turning a crank attached to the machine, a constant current of galvanic electricity; and as there is no consumption of material necessary to obtain this power, it will doubtless supersede the use of the galvanic battery, which, in the event of constant employment, would be very expensive, from the waste of zinc, platinum, acids, mercury, and other materials used in its construction. It particularly recommends itself for magnetizing purposes, as it requires no knowledge of chemistry to insure the result, being merely mechanical in its action, and is always ready for action without previous preparation; the turning of a crank being the only requisite when the machine is in order. It is not liable to get out of order; does not diminish perceptibly in power when in constant use, and actually gains power when standing at rest. It will be particularly gratifying to the man of science, as it enables him to have always at hand a constant power for the investigation of its properties, without any labor of preparation. We notice among the beautiful results of this machine, that it charges an electro magnet so as to sustain a weight of 1,000 pounds, and it ignites to a white heat large platinum wires, and may be used successfully for blasting at a distance; and should Government ever adopt any such system of defence as to need the galvanic power, it must supersede the battery in that case. Professor Page demonstrates, by mathematical reasoning, that the new contrivance of the coils affords the very maximum of quantity to be obtained by magnetic excitation.

Report of Commissioner of Patents, for 1844.

REPORTS TO CONGRESS ON THE SUBJECT OF ELECTRO MAGNETIC TELEGRAPHS.

Letter from the Secretary of the Treasury, transmitting a Report upon the subject of a System of Telegraphs for the United States. December 11, 1837.

TREASURY DEPARTMENT, *December 6, 1837.*

SIR: I have the honor to present this report, in compliance with the following resolution, which passed the House of Representatives on the 3d of February last, viz. "*Resolved*, That the Secretary of the Treasury be requested to report to the House of Representatives, at its next session, upon the propriety of establishing a system of telegraphs for the United States." Immediately after its passage I prepared a circular, with the view of procuring, from the most intelligent sources, such information as would enable Congress, as well as the Department, to decide upon the propriety of establishing a system of telegraphs.

It seemed also important to unite with the inquiry the procurement of such facts as might show the expense attending different systems; the celerity of communication by each; and the useful objects to be accomplished by their adoption.

A copy of the circular is annexed, (1)

The replies have been numerous, and many of them are very full and interesting. Those deemed material are annexed, numbered 2 to 18, inclusive.

From those communications, and such other investigations as the pressure of business has enabled me to make, I am satisfied that the establishment of a system of telegraphs for the United States would be useful to commerce as well as the Government. It might most properly be made appurtenant to the Post Office Department, and, during war, would prove a most essential aid to the military operations of the country.

The expense, attending it is estimated carefully in some of the documents annexed; but it will depend much upon the kind of system adopted: upon the extent and location of the lines first established; and the charges made to individuals for communicating information through it which may not be of a public character.

On these points, as the Department has not been requested to make a report, no opinion is expressed; but information concerning them was deemed useful as a guide in deciding on the propriety of establishing telegraphs, and was, therefore, requested in the circular before mentioned. Many useful suggestions in relation to the subject will be found in the correspondence annexed, and in the books there referred to.

The Department would take this occasion to express, in respect to the numerous gentlemen whose views are now submitted to Congress, its high appreciation and sincere acknowledgments for the valuable contributions they have made on a subject of so much interest.

I remain, very respectfully,
Your obedient servant,

LEVI WOODBURY,
Secretary of the Treasury.

The Hon. J. K. POLK,
Speaker of the House of Representatives.

No. 1.

Circular to certain Collectors of the Customs, Commanders of Revenue Cutters, and other persons.

TREASURY DEPARTMENT, *March 10, 1837.*

With the view of obtaining information in regard to "the propriety of establishing a system of telegraphs for the United States," in compliance with the request contained in the annexed resolution of the House of Representatives, adopted at its last session, I will thank you to furnish the Department with your opinion upon the subject. If leisure permits, you would oblige me by pointing out the manner, and the various particulars, in which the system may be rendered most useful to the Government of the United States and the public generally. It would be desirable, if in your power, to present a detailed statement as to the proper points for the location, and distance of the stations from each other, with general rules for the regulation of the system, together with your sentiments as to the propriety of connecting it with any existing department of the Government, and some definite idea of the rapidity with which intelligence could ordinarily, and also in urgent cases, be communicated between distant places. I wish you to estimate the probable expense of establishing and supporting telegraphs, upon the most approved system, for any given distance, during any specified period.

It would add to the interest of the subject if you would offer views as to the practicability of uniting with a system of telegraphs for communication in clear weather and in the day time, another for communication in fogs, by cannon, or otherwise; and in the night, by the same mode, or by rockets, fires, &c.

I should be gratified by receiving your reply by the first of October next.

LEVI WOODBURY,
Secretary of the Treasury.

No. 2.

Letter from S. F. B. Morse, to the Secretary of the Treasury.

NEW YORK CITY UNIVERSITY, *Sept. 27, 1837.*

DEAR SIR: In reply to the inquiries which you have done me the honor to make, in asking my opinion "of the propriety of establishing a system of telegraphs for the United States," I would say, in regard to the general question, that I believe there can scarcely be two opinions, in such a community as ours, in regard to the advantage which would result, both to the Government and the public generally, from the establishment of a system of communication by which the most speedy intercourse may be had between the most distant parts of the country. The *mail system*, it seems to me, is founded on the universally admitted principle, that the greater the speed with which intelligence can be transmitted from point to point, the greater is the benefit derived to the whole community. The only question that remains, therefore, is, what system is best calculated, from its completeness and cheapness, to effect this desirable end?

With regard to telegraphs constructed on the ordinary principles, however perfected within the limits in which they are necessarily confined, the most perfect of them are liable to one insurmountable objection—*they are useless the greater part of the time.* In foggy weather, and ordinarily during the night no intelligence can be transmitted. Even when they can transmit, much time is consumed in communicating but little, and that little not always precise.

Having invented an entirely new mode of telegraphic communication, which, so far as experiments have yet been made with it, promises results of almost marvellous character, I beg leave to present to the Department a brief account of its chief characteristics.

About five years ago, on my voyage home from Europe, the electrical experiment of Franklin, upon a wire some four miles in length was casually recalled to my mind in a conversation with one of the passengers, in which experiment it was ascertained that the electricity travelled through the whole circuit in a time not appreciable, but apparently instantaneous. *It immediately occurred to me, that if the presence of electricity could be made* VISIBLE *in any desired part of this circuit, it would not be difficult to construct a* SYSTEM OF SIGNS *by which intelligence could be instantaneously transmitted.* The thought, thus conceived, took strong hold of my mind in the leisure which the voyage afforded, and I planned a system of signs and an apparatus to carry it into effect. I cast a species of type, which I had devised for this purpose, the first week after my arrival home; and although the rest of the machinery was planned, yet, from the pressure of unavoidable duties, I was compelled to postpone my experiments, and was not able to test the whole plan until within a few weeks. The result has realized my most sanguine expectations.

As I have contracted with Mr. Alfred Vail to have a complete apparatus made to demonstrate at Washington by the 1st of January, 1838, the practicability and superiority of my mode of telegraphic communication by means of electro magnetism, (an apparatus which I hope to have the pleasure of exhibiting to you,) I will confine myself in this communication to a statement of its peculiar advantages.

First. The *fullest and most precise information* can be almost instantaneously transmitted between any two or more points, between which a wire conductor is laid: that is to say, no other time is consumed than is necessary to write the intelligence to be conveyed, and to convert the words into the telegraphic numbers. The numbers are then transmitted nearly instantaneously, (or if I have been rightly informed in regard to some recent experiments in the velocity of electricity, *two hundred thousand miles in a second,*) to any distance, where the numbers are immediately recognised, and reconverted into the words of the intelligence.

Second. The same full intelligence can be communicated *at any moment irrespective of the time of day or night, or state of the weather.* This single point establishes its superiority to all other modes of telegraphic communication now known.

Third. The whole apparatus will occupy but *little space,* (scarcely six cubic feet, probably not more than four;)[11] and it may, therefore, be placed without inconvenience, in any house.

Fourth. The *record of intelligence is made in a permanent manner and in such a form* that it can be at once bound up in *volumes* convenient for reference, if desired.

Fifth. Communications are secret to all but the persons for whom they are intended.

These are the chief advantages of the Electro Magnetic Telegraph over other kinds of telegraphs, and which must give it the preference, provided the expense and other circumstances are reasonably favorable.

The newness of the whole plan makes it not so easy to estimate the expense, but an *approach* to a correct estimate can be made.

The principal expense will be the first cost of the wire or metallic conductors, (consisting of four lengths,) and the securing them against injury. The cost of a single copper wire $\frac{1}{16}$ of an inch diameter, (and it should not be of less dimensions,) for 400 miles, was recently estimated in Scotland to be about

£1,000 sterling, including the solderings of the wire together; that is, about $6 per mile for one wire, or $24 per mile for the four wires. I have recently contracted for twenty miles of copper wire, No. 18, at 40 cents per pound. Each pound, it is estimated, contains 93 feet, which gives a result coinciding with the Scotch estimate, if $1.60 per mile be added for solderings.

The preparation of the wire for being laid, (if in the ground,) comprehends the *clothing of the wires* with an insulating or non-conducting substance; the *encasing them in wood, clay, stone, iron*, or *other metal*; and the *trenching* of the earth to receive them. In this part of the business I have no experience to guide me, the whole being altogether new. I can, therefore, only make at present a rough estimate. Iron tubes enclosing the wires, and filled in with pitch and resin, would probably be the most eligible mode of securing the conductors from injury, while, at the same time, it would be the most costly. Iron tubes of 1½ inch diameter, I learn, can be obtained at Baltimore, at 28 cents per foot. The *trenching* will not be more than three cents for 2 feet, or about $75 per mile. This estimate is for a trench 3 feet deep and 1½ wide. There is no *grading*; the trench may follow the track of any road, over the highest hills or lowest valleys. Across rivers, with bridges, the circuit may easily be carried, enclosed beneath the bridge. Where the stream, is wide, and no bridge, the circuit, enclosed in lead, may be sunk to the bottom.

If the circuit is laid through the air, the first cost would doubtless be much lessened. This plan of making the circuit has some advantages, but there are also some disadvantages; the chief of which latter is, that, being always in sight, the temptation to injure the circuit to mischievously disposed persons, is greater than if it were buried out of sight beneath their feet. As an offset, however, to this, an injury to the circuit is more easily detected. With regard to danger from wantonness, it may be sufficient to say, that the same objection was originally made in the several cases, successively, of water-pipes, gas-pipes, and railroads; and yet we do not hear of wantonness injuring any of these. Stout spars of some thirty feet in height, well planted in the ground, and placed about 350 feet apart, would, in this case, be required, along the tops of which the circuit might be stretched. Fifteen such spars would be wanted to a mile. This mode would be as cheap, probably, as any other, unless the laying of the circuit in water should be found to be most eligible. A series of experiments to ascertain the practicability of this mode, I am about to commence with Professor Gale, of our university, a gentleman of great science, and to whose assistance, in many of my late experiments, I am greatly indebted. We are preparing a circuit of twenty miles. The result of our experiments I will have the honor of reporting to you.

The other machinery, consisting of the apparatus for transmitting and receiving the intelligence, can be made at a very trifling cost. The only parts of the apparatus that waste or consume materials, are the batteries, which consume *acid* and *zinc*, and the register, which consumes *paper* for recording, and *pencils* or *ink* for marking.

The cost of *printing*, in the first instance, of a *telegraphic dictionary*,[12] should perhaps also be taken into the account, as each officer of the Government, as well as many others, would require a copy, should this mode of telegraphic communication go into effect. This dictionary would contain a vocabulary of all the words in common use in the English language, with the numbers regularly affixed to each word.

The stations in the case of this telegraph may be as numerous as are desired; the only additional expense for that purpose being the adding of the transmitting and receiving apparatus to each station.

The cost of supporting a system of telegraphs on this plan, (when a circuit is once established,) would, in my opinion, be much less than on the common plans; yet, for want of experience in this mode, I would not affirm it positively.

As to "the propriety of connecting the system of telegraphs with any existing department of Government," it would seem most natural to connect a telegraphic system with the Post Office Department; for, although it does not carry a mail, yet it is another mode of accomplishing the principal object for which the mail is established, to wit: the rapid and regular transmission of intelligence. If my system of telegraphs should be established, it is evident that the telegraph would have but little rest day or night. The advantage of communicating intelligence instantaneously in hundreds of instances of daily occurrence, would warrant such a rate of *postage*, (if it may be so called,) as would amply defray all expenses of the first cost of establishing the system, and of guarding it, and keeping it in repair.

As every word is numbered, an obvious mode of rating might be, a *charge of a certain amount on so many numbers*. I presume that five words can certainly be transmitted in a minute; for, with the imperfect machinery I now use, I have recorded at that rate, at the distance of half a mile.

In conclusion, I would say, that if the perfecting of this new system of telegraphs (which may justly be called the American Telegraph, since I can establish my claims to priority in the invention) shall be thought of public utility, and worthy the attention of Government, I shall be ready to make any sacrifice of personal service and of time to aid in its accomplishment.

In the mean time I remain, sir, with sincere respect and high personal esteem,

Your most obedient, humble servant,
SAML. F. B. MORSE.

HON. LEVI WOODBURY,
Secretary of the Treasury.

No. 3.
Letter from S. F. B. Morse to the Secretary of the Treasury.

UNIVERSITY OF THE CITY OF NEW YORK,
November 28, 1837.

MY DEAR SIR: In my letter to you in answer to the circular respecting telegraphs, which you did me the honor to send me, I promised to advise you of the result of some experiments about to be tried with my electro magnetic telegraph. I informed you that I had succeeded in marking permanently and intelligibly at the distance of *half a mile.*

Professor Gale, of our university, and Mr. Alfred Vail, of the Speedwell iron-works, near Morristown, New Jersey, are now associated with me in the scientific and mechanical parts of the invention. We have procured several miles of wire, and I am happy to announce to you that our success has thus far, been complete. At a distance of *five miles*, with a common Cruikshank's battery of 87 plates, (4 by 3½ inches each plate,) the marking was as perfect on the register as in the first instance of half a mile. We have recently added *five miles more*, making in all *ten miles*, with the *same result*; and we have now no doubt of its effecting a *similar result* at *any distance.*

I also stated to you, sir, that machinery was in progress of making, with which, so soon as it should be completed, I intended to proceed to Washington, to exhibit the powers of the invention before you and other members of the Government. I had hoped to be in Washington before the session of Congress, but I find that the execution of new machinery is so uncertain in its time of completion, that I shall be delayed, probably, until the beginning of the year.

What I wish to learn from you, sir, is, "*How late in the session can I delay my visit, and yet be in season to meet the subject of telegraphs, when it shall be presented by your report to Congress?*"

I am anxious, of course, to show as perfect an instrument as possible, and would wish as much time for the purpose of perfecting it as can be allowed without detriment to my interests as an applicant for the attention of Government to the best plan of a telegraph.

I am, my dear sir, with the greatest respect and personal esteem,

Your most obedient servant,
SAML. F. B. MORSE.

HON. LEVI WOODBURY,
Secretary of the Treasury.

No. 4.
[From the New York Journal of Commerce.]

We have received the following note and diagram, with the explanation of the latter, from Mr. Morse:

To the Editors of the Journal of Commerce:

GENTLEMEN: You had the kindness to assert, a few days ago, my claim to the invention of the *electro magnetic telegraph*, for which I thank you. As to the priority of my invention, entirely planned and for the most part executed as it was nearly five years ago, I can adduce the amplest proof.

You announced that I was preparing a short *circuit*, to show to my friends the operation of the telegraph. This circuit I have completed, of the length of 1,700 feet, or about one-third of a mile; and on Saturday, the 2d instant, in presence of Professors Gale and Torrey of this city, and Professor Daubeny of the Oxford (English) University, and several other gentlemen, I tried a preliminary experiment with the register. It recorded the intelligence sufficiently perfect to establish the practicability of the plan, and the superior simplicity of my mode of communication, over any of those proposed by the professors in Europe.

It will be observed that no account has reached us that any of the foreign proposed electric telegraphs have as yet succeeded in transmitting intelligible communications; but it is merely asserted of the most advanced experiment, (the one in London,) that "by means of five wires," &c. intelligence "*may be* conveyed." I have the gratification of sending you a specimen of the writing of my telegraph, the actual transmission of a communication made this morning, in a more complete manner than on Saturday, and through the distance of one-third of a mile.

Thinking it may be gratifying to your readers to see the kind of writing which it performs, I have had it engraved for you, accompanied with an explanation.

Your obedient servant,
SAML. F. B. MORSE.

N. Y. City University, September 4, 1837.

No. 5.

Specimen of Telegraphic Writing made by means of electricity at the distance of one-third of a mile.

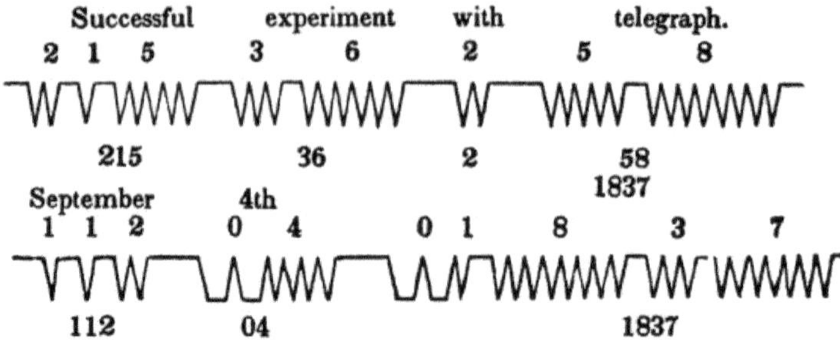

The *words* in the diagram were the intelligence transmitted.

The *numbers* (in this instance arbitrary) are the numbers of the words in a telegraphic dictionary.

The *points* are the markings of the register, each point being marked every time the electric fluid passes.

The register marks but one kind of mark, to wit, (**V**.) This can be varied two ways. By intervals, thus, (**V VV VVV**,) signifying one, two, three, &c., and by reversing, thus, (**Λ**) Examples of both these varieties are seen in the diagram.

The single numbers are separated by short, and the whole numbers by *long intervals*.

To illustrate by the diagram: the word "successful" is first found in the dictionary, and its telegraphic number, 215, is set up in a species of type prepared for the purpose, and so of the other words. The type then operate upon the machinery, and serve to regulate the times and intervals of the passage of electricity. Each passage of the fluid causes a pencil at the extremity of the wire to mark the points as in the diagram.

To read the marks, count the points at the bottom of each line. It will be perceived that two points come first, separated by a *short* interval from the next point. Set 2 beneath it. Then comes one point, likewise separated by a *short* interval. Set 1 beneath it. Then come five points. Set 5 beneath them. But the interval in this case is a *long* interval: consequently, the three numbers comprise the whole number, 215.

So proceed with the rest, until the numbers are all set down. Then, by referring to the telegraphic dictionary, the words corresponding to the numbers are found, and the communication read. Thus it will be seen that, by means of the changes upon *ten* characters, all words can be transmitted.

But there are *two points* reversed in the lower line. These are the *eleventh* character, placed before a number, to signify that it is to be read as a *number*, and not as the representative of a word.

No. 6.

Mr. SMITH, from the Committee on Commerce, made the following Report, April 6th, 1838.

The Committee on Commerce to whom the subject was referred, have had the same under consideration, and report:

On the 3d of February, 1837, the House of Representatives passed a resolution requesting the Secretary of the Treasury to report to the House, at its present session, upon the propriety of establishing a system of telegraphs for the United States.

In pursuance of this request, the Secretary of the Treasury, at an early day after the passage of said resolution, addressed a circular of inquiry to numerous scientific and practical individuals in different parts of the Union; and, on the 6th of December last, reported the result of this proceeding to the House.

This report of the Secretary imbodies many useful suggestions on the necessity and practicability of a system of telegraphic despatches, both for public and individual purposes; and the committee cannot doubt that the American public is fully prepared, and even desirous, that every requisite effort be made on the part of Congress to consummate an object of so deep interest to the purposes of Government in peace and in war, and to the enterprise of the age.

Amid the suggestions thus elicited from various sources, and imbodied in the before-mentioned report of the Secretary of the Treasury, a plan for an electro magnetic telegraph is communicated by Professor Morse, of the University of the city of New York, pre-eminently interesting, and even wonderful. See Report, No. 2.

This invention consists in the application, by mechanism, of galvanic electricity to telegraphic purposes, and is claimed by Professor Morse and his associates as original with them; and being so, in fact, as the committee believe, letters-patent have been secured, under the authority of the United States, for the invention. It has, moreover, been subjected to the test of experiment, upon a scale of ten miles distance, by a select committee of the Franklin Institute of the city of Philadelphia, and reported upon by that eminently high tribunal in the most favorable and confident terms. An abstract from the report thus made is hereunto annexed. No. 7.

In additional confirmation of the merits of his proposed system of telegraphs, Professor Morse has exhibited it in operation (by a coil of metallic wire measuring about ten miles in length, rendering the action equal to a telegraph of half that distance) to the Committee on Commerce of the House of Representatives, to the President of the United States, and the several heads of Departments, to members of Congress generally, who have taken interest in the examination, and to a vast number of scientific and practical individuals from various parts of the Union; and all concur, it is believed, and without a dissenting doubt, in admiration of the ingenious and scientific character of the invention, and in the opinion that it is successfully adapted to the purposes of telegraphic despatches, and in a conviction of its great and incalculable practical importance and usefulness to the country, and ultimately to the whole world.

But it would be presumptuous in any one, (and the inventor himself is most sensible of this,) to attempt, at this stage of the invention, to calculate in anticipation, or to hold out promises of what its whole extent of capacity for usefulness may be, in either a political, commercial, or social point of view, if the electrical power upon which it depends for successful action shall prove to be efficient, as is now supposed it will, to carry intelligence through any of the distances of 50, 100, 500, or more miles now contemplated. No such attempt, therefore, will be indulged in this report. It is obvious, however, that the influence of this invention over the political, commercial, and social relations of the people of this widely-extended country, looking to nothing beyond, will, in the event of success, of itself amount to a revolution unsurpassed in moral grandeur by any discovery that has been made in the arts and sciences, from the most distant period to which authentic history extends, to the present day. With the means of almost instantaneous communication of intelligence between the most distant points of the country, and simultaneously between any given number of intermediate points which this invention contemplates, space will be, to all practical purposes of information, completely annihilated between the States of the Union, as also between the individual citizens thereof. The citizen will be invested with, and reduce to daily and familiar use, an approach to the HIGH ATTRIBUTE OF UBIQUITY, in a degree that the human mind until recently, has hardly dared to contemplate seriously as belonging to human agency, from an instinctive feeling of religious reverence and reserve on a power of such awful grandeur.

Referring to the annexed report of the Franklin Institute, already adverted to, and also to the letters of Professor Morse, marked 2, 8 and 9, for other details of the superiority of this system of telegraphs over all other methods heretofore reduced to practice by any individual or Government, the committee agree, unanimously, that it is worthy to engross the attention and means of the Federal Government, to the full extent that may be necessary to put the invention to the most decisive test that can be desirable. The power of the invention, if successful, is so extensive for good and for evil, that the Government alone should possess the right to control and regulate it. The mode of proceeding to test it, as suggested, as also the relations which the inventor and his associates are willing to recognise with the Government on the subject of the future ownership, use, and control of the invention, are succinctly set forth in the annexed letters of Professor Morse, marked 8 and 9.

The probable outlay of an experiment upon a scale equal to fifty miles of telegraph, and equal to a circuit of double that distance, is estimated at $30,000. Two-thirds of this expenditure will be for material, which, whether the experiment shall succeed or fail, will remain uninjured, and of very little diminished value below the price that will be paid for it.

The estimates of Professor Morse, as will be seen by his letter, marked 9, amount to $26,000; but, to meet any contingency not now anticipated, and to guard against any want of requisite funds in an enterprise of such moment to the Government, to the people, and to the scientific world, the committee recommend an appropriation of $30,000, to be expended under the direction of the Secretary of the Treasury; and to this end submit herewith a bill.

It is believed by the committee that the subject is one of such universal interest and importance, that an early action upon it will be deemed desirable by Congress, to enable the inventor to complete his trial of the invention upon the extended scale contemplated, in season to furnish Congress with a full report of the result during its present session, if that shall be practicable.

All which is respectfully submitted.

FRANCIS O. J. SMITH, JAS. M. MASON,

S. C. PHILLIPS, JOHN T. H. WORTHINGTON,

SAMUEL CUSHMAN, WM. H. HUNTER,

JOHN I. DE GRAFF, GEORGE W. TOLAND,

EDWARD CURTIS,

Committee on Commerce, U. S. H. R.

No. 7.

HALL OF THE FRANKLIN INSTITUTE, *Feb. 8, 1838.*

Report of the Franklin Institute Philadelphia.

The sub-committee, from the committee of science and arts, appointed to examine the electro magnetic telegraph of Professor Samuel F. B. Morse, report:

That this instrument was exhibited to them in the hall of the Institute, and every opportunity given by Mr. Morse and his associate, Mr. Alfred Vail, to examine it carefully, and to judge of its operation; and they now present the following as the result of their observations:

* * * * The operation of the telegraph, as exhibited to us, was very satisfactory. The power given to the magnet at the register, through a length of wire of ten miles, was abundantly sufficient for the movements required to mark the signals. The communication of this was instantaneous. The time required to make the signals was as short at least, as that necessary in the ordinary telegraphs. It appears to this committee, therefore, that the possibility of using telegraphs upon this plan, in actual practice, is not to be doubted; though difficulties may be anticipated, which could not be tested by the trials made with the model.

One of these relates to the insulation and protection of the wires, which are to pass over many miles of distance, to form the circuits between the stations. Mr. Morse has proposed several plans: the last being to cover the wires with cotton thread, then varnish them thickly with gum-elastic, and enclose the whole in leaden tubes. More practical and economical means will probably be devised; but the fact is not to be concealed, that any effectual plan must be very expensive.

Doubts have been raised as to the distance to which the electricity of an ordinary battery can be made efficient; but your committee think that no serious difficulty is anticipated as to this point. The experiment with the wire wound in a coil, may not, indeed, be deemed conclusive; but one of the members of the committee assisted in an experiment in which a magnet was very sensibly affected by a battery of a single pair, through an insulated wire of two and three-quarter miles in length, of which the folds were four inches apart; and when a battery of ten pairs was used, water was freely decomposed. An experiment is said to have been made, with success, on the Birmingham and Manchester rail road, through a circuit of thirty miles in length.

It may be proper to state, that the idea of using electricity for telegraphic purposes has presented itself to several individuals, and that it may be difficult to settle among them the question of originality. The celebrated

Gauss has a telegraph of this kind in actual operation, for communicating signals between the University of Göttingen and his magnetic observatory in its vicinity. Mr. Wheatstone, of London, has been for some time also engaged in experiments on an electrical telegraph. But the plan of Professor Morse is, so far as the committee are informed, entirely different from any of those devised by other individuals, all of which act by giving different *directions* to a magnetic needle; and would, therefore, require several circuits of wire between all the stations.

In conclusion, the committee beg leave to state their high gratification with the exhibition of Professor Morse's telegraph, and their hope that means may be given to him to subject it to the test of an actual experiment, made between stations at a considerable distance from each other. The advantages which this telegraph would present, if successful, over every kind heretofore used, make it worthy the patronage of the Government. These are, that the stations may be at a distance asunder, far exceeding that to which all other telegraphs are limited; and that the signals may be given at night, and in rains, snows, and fogs, when other telegraphs fail.

R. M. PATTERSON, *Chairman.*

No. 8.

From S. F. B, Morse, to the Hon. F. O. J. Smith.

WASHINGTON, *February 15, 1838.*

DEAR SIR: In consequence of the conversation had with the committee on the subject of my telegraph, I would state, that I think it desirable that an experiment, on a somewhat extended scale, should first be made to test both the practicability and the facility of communicating intelligence for at least one hundred miles. The experiment may proceed, as to cost, with perfect safety to the Government. *First.* The wire for this distance, consisting of four lengths, making a total of four hundred miles of wire, might be obtained, and receive its covering of cotton and other insulation. This length would amply suffice to ascertain the law of the propulsive power of voltaic electricity, and previous to any measures being taken for burying it in the earth. So that, if any unforeseen difficulty should occur fatal to its practicability, the wire is not consumed or lost. If the expected success is realized, then, *Second.* The preparation of the wire might be commenced for burying in the earth, and, being found complete through the whole route, the several portrules, registers, batteries, &c., might be provided to put the telegraph into complete action. This experiment of one hundred miles would furnish the data from which to make the estimates of a more general extension of the system. If no insurmountable obstacles present themselves in a distance of one hundred miles, none may be expected in one thousand or in ten thousand miles; and then will be presented for the consideration of the Government the propriety of completely organizing the new telegraphic system as a part of the Government, attaching it to some department already existing, or creating a new one, which may be called for by the accumulating duties of the present departments.

It is obvious, at the slightest glance, that this mode of instantaneous communication must inevitably become an instrument of immense power, to be wielded for good or for evil, as it shall be properly or improperly directed. In the hands of a company of speculators, who should monopolize it for themselves, it might be the means of enriching the corporation at the expense of the bankruptcy of thousands; and even in the hands of Government alone, it might become a means of working vast mischief to the republic. In considering these prospective evils, I would respectfully suggest a remedy which offers itself to my mind. Let the sole right of using the telegraph belong, in the first place, to the Government, who should grant, for a specified sum or bonus, to any individual or company of individuals who may apply for it, and under such restrictions and regulations as the Government may think proper, the right to lay down a way communication between any two points, for the purpose of transmitting intelligence; and thus would be promoted a general competition. The Government would have a telegraph of its own, and have its modes of communicating with its own officers and agents, independent of private permission, or interference with and interruption to the ordinary transmissions on the private telegraphs. Thus there would be a system of checks and preventives of abuse, operating to restrain the action of this otherwise dangerous power, within those bounds which will permit only the good and neutralize the evil. Should the Government thus take the telegraph solely under its own control, the revenue derived from the bonuses alone, it must be plain, will be of vast amount. From the enterprising character of our countrymen, shown in the manner in which they carry forward any new project which promises private or public advantage, it is not visionary to suppose that it would not be long ere the whole surface of this country would be channelled for those *nerves* which are to diffuse, with the speed of thought, a knowledge of all that is occurring throughout the land; making, in fact, *one neighborhood* of the whole country.

If the Government is disposed to test this mode of telegraphic communication by enabling me to give it a fair trial for one hundred miles, I will engage to enter into no arrangements to dispose of my rights as the inventor and patentee for the United States, to any individual or company of individuals, previous to offering it to the Government for such a just and reasonable compensation as shall be mutually agreed upon.[13]

I remain, sir, respectfully, your most obedient servant,
SAMUEL F. B. MORSE.

To the Hon. F. O. J. SMITH,
Chairman of the Committee on Commerce
of the House of Representatives.

No. 9.

Letter from S. F. B. Morse to Hon. F. O. J. Smith.

WASHINGTON, *February 22, 1838.*

DEAR SIR: I have endeavoured to approach a proper estimate of the expense attendant on preparing a complete telegraphic communication for some distance; and taking into consideration the possibility that the experiment may be conclusively tried before the close of the present session of Congress, I have thought that an appropriation for fifty miles of distance would test the practicability of the telegraph quite as satisfactorily as one hundred, because the obstacles necessary to be overcome would not be more proportionally in fifty than in one hundred; while, at the same time, the *double circuit* necessary in the fifty miles would give a *single circuit* of one hundred for the purpose of testing the effect of distance upon the passage of electricity. Fifty miles would require a less amount of appropriation, and the experiment could also be sooner brought to a result.

Two hundred miles of wire, or wire for two circuits for fifty miles of distance, including the covering of the wire with cotton, at $100 per mile,	$20,000
Other expenses of preparation of the wire, such as caoutchouc, wax, resin, tar, with reels for winding, soldering, &c., say $6 per mile,	1,200
Batteries and registers, with type, &c., for two stations, and materials for experimenting on the best modes of magnets at long distances,	800
Services of Professor Gale in the chemical department; services of Mr. Alfred Vail in the mechanical department; services of assistants in different departments; my own services in superintending and directing the whole—total	4,000
[14]	
Total,	$26,000

This estimate is exclusive of expense necessary to lay down the wire beneath the ground. This is unnecessary until the previous preparations are found satisfactory.

I cannot say what time will be required for the completion of the circuits for fifty miles. If the order could be immediately given for the wire, I think all the other matter connected with it might be completed so that every thing would be in readiness in *three months*. Much will depend on the punctuality with which contractors fulfil their engagements in furnishing the wire and other apparatus.

I remain, sir, very respectfully, your obedient servant,
SAMUEL F. B. MORSE.

To the Hon. F. O. J. SMITH,
Chairman of the Committee on Commerce.
of the House of Representatives.

No. 10.

Mr. Ferris, from the Committee on Commerce, made the following Report, December 30, 1842.

That they regard the question, as to the general utility of the telegraphic system, settled by its adoption by the most civilized nations; and experience has fully demonstrated the great advantages which may be derived from its use. Its capability of speedily transmitting intelligence to great distances, for national defence, and for other purposes, where celerity is desirable, is decidedly superior to any of the ordinary modes of communication in use. By it, the first warning of approaching danger, and the appearance of hostile fleets and armies on our coasts and borders, may be announced simultaneously at the most distant points of our widely-extended empire, thus affording time and opportunity for concentrating the military force of the country, for facilitating military and naval movements, and for transmitting orders suitable to the emergency.

In the commercial and social affairs of the community, occasions frequently arise in which the speedy transmission of intelligence may be of the highest importance for the regulation of business transactions, and in relieving the anxious solicitude of friends, as to the health and condition of those in whose fortunes they feel an interest.

The practicability of establishing telegraphs on the electric principle is no longer a question. Wheatstone, of London, and his associates, have been more fortunate than our American inventor, in procuring the means to put his ingenious system into practical use for two or three hundred miles, in

Great Britain; and the movements of the cars on the Blackwall rail road are at this time directed with great economy, and perfect safety to life and property, by means of his magnetic needle telegraph. If a system more complicated and less efficient than the American telegraph is operated for great distances in England, with such eminent success and advantage, there can be no reasonable doubt that, if the means be furnished for putting in operation the system of Professor Samuel F. B. Morse, of New York, the original inventor of the electro-magnetic telegraph, the same, if not greater success, will be the result. Your committee are of opinion that it is but justice to Professor Morse, who is alike distinguished for his attainments in science and excellence in the arts of design, and who has patiently devoted many years of unremitting study, and freely spent his private fortune, in inventing and bringing to perfection a system of telegraphs which is calculated to advance the scientific reputation of the country, and to be eminently useful, both to the Government and the people, that he should be furnished with the means of competing with his European rivals.

Professor Morse bases his system upon the two following facts in science:

First. That a current of electricity will pass to any distance along a conductor connecting the two poles of a voltaic battery or generator of electricity, and produce visible effects at any desired points on that conductor.

Second. That magnetism is produced in a piece of soft iron (around which the conductor, in its progress, is made to pass) when the electric current is permitted to flow, and that the magnetism ceases when the current of electricity is prevented from flowing. This current of electricity is produced and destroyed by breaking and closing the galvanic circuit at the pleasure of the operator of the telegraph, who in this manner directs and controls the operation of a simple and compact piece of mechanism, styled the register, which, at the will of the operator at the point of communication, is made to record, at the point of reception, legible characters, on a roll of paper put in motion at the same time with the writing instrument. These characters the inventor has arranged into a conventional *alphabet*, and which is capable of being learned and used with very little practice.

Professor Morse has submitted his telegraphic plan to the severe scrutiny of European criticism; and the Academy of Sciences, of Paris, the highest scientific tribunal in the world, hailed it with enthusiasm and approbation, when its operation was exhibited, and its principles explained by their distinguished perpetual secretary, M. Arago.

It appears, from documents produced by Professor Morse, that the thanks of several learned bodies in France were voted to him for his invention, and the large medal of honor was awarded to him by the Academy of Industry. It further appears, that several other systems of telegraphs on the electric plan (among which were Wheatstone's, of London, Steinheil's, of Munich, and Masson's, of Caen) had been submitted at various times for the consideration of the French Government, who appointed a commission to examine and report on them all, at the head of which commission was placed the administrator-in-chief of the telegraphs of France, (M. Foy,) who, to a note to Professor Morse, thus writes:

"I take a true pleasure in confirming to you in writing that which I have already had the honor to say to you viva voce—that I have prominently presented to Monsieur the Minister of the Interior your electro magnetic telegraph, as being the system which presents the best chance of a practical application; and I have declared to him that, if some trials are to be made with electric telegraphs, I do not hesitate to recommend that they should be made with your apparatus."

Your committee, in producing further evidence of the approbation by the scientific world of the system of Professor Morse, would cite the letter of Professor Henry, of Princeton College, well known for his eminent attainments in electrical science, (marked 11,) in the appendix of this report.

More recently, a committee, consisting of some of our most distinguished scientific citizens, was appointed by the American Institute of New York, to examine and report upon this telegraph, who made the report (12) in the appendix. In compliance with the recommendation of this report, the Institute awarded to Professor Morse the gold medal.

Besides the evidence these testimonials furnish of the excellence of Professor Morse's system, your committee, as well as the greater part of the members of both Houses of Congress, have had a practical demonstration of the operation of the electro magnetic telegraph, and have witnessed the perfect facility and extraordinary rapidity with which a message can be sent by means of it from one extremity of the capitol to the other. This rapidity is not confined in its effects to a few hundred feet, but science makes it certain that the same effects can be produced, at any distance on the globe, between any two given points connected by the conductors.

Your committee have alluded to other electric telegraphs; for, as is not uncommon in the birth of great inventions, scientific minds have, at nearly the same period of time, in various parts of Europe, conceived and planned electric telegraphs; but it is a matter of national pride, that the invention of the *first electro magnetic telegraph*, by Professor Morse, as well as the *first conception* of using electricity as the means of transmitting intelligence, by Doctor Franklin, is the offspring of American genius.

Your committee beg leave to refer to the letter of Professor Morse, (marked 13,) in the appendix, to C. G. Ferris, one of the committee, giving, at his request a brief history of the telegraph since it was before Congress, in 1838, for some interesting information concerning it, and for Professor Morse's estimate of the probable expense of establishing his system of telegraphs for thirty or forty miles.

They would also refer to the House document, No. 15, (December 6, 1837,) and to House report, No. 753, (April 6, 1838,) for valuable information on the subject of telegraphs.

Your committee invite special attention to that part of Professor Morse's letter which details the plan of a *revenue* which may be derived from his telegraphic system, when established to an extent sufficient for the purposes of commercial and general intelligence. From these calculations, made upon safe data, it is probable that an income would be derived from its use by merchants and citizens more than sufficient to defray the interest of the capital expended in its establishment. So inviting, indeed, are the prospects of profit to individual enterprise, that it is a matter of serious consideration, whether the Government should not, on this account alone, seize the present opportunity of securing to itself the regulation of a system which, if monopolized by a private company, might be used to the serious injury of the Post Office Department, and which could not be prevented without such an interference with the rights of the inventor and of the stockholders as could not be sustained by justice or public opinion.

After the ordeal to which the electro magnetic telegraph system has been subjected, both in Europe and in America, and the voice of the scientific world in its favor, it is scarcely necessary for your committee to say that they have the fullest confidence in Professor Morse's plan, and they earnestly recommend the adoption of it by the Government of the United States. They deem it most fortunate that no definite system of telegraphs should hitherto have been adopted by the Government, since it enables them to establish this improved system, which, in the opinion of your committee, is decidedly superior to any other now in use, possessing an advantage over telegraphs depending on vision, inasmuch as it may be used both by night and day, in all weathers, and in all seasons of the year, with equal convenience; and, also, possessing an advantage over electric telegraphs heretofore in use, inasmuch as it records, in permanent legible characters on paper, any communication which may be made by it, without the aid of any agent at the place of recording, except the apparatus which is put in motion at the point of communication. Thus, the recording apparatus, called the register, may be left in a closed chamber, where it will give notice of its commencing to write, by a bell, and the communication may be found, on opening the apartment. Possessing these great advantages, and the means of communication not being liable to interruption by the ordinary contingencies which may impede or prevent the successful action of other telegraphs, the advantages to be derived from it will soon be apparent to the community, and it will become the successful rival of the Post Office, when celerity of communication is desired, and create a revenue from which this system of telegraphs may be extended and ramified through all parts of the country, without imposing any burden upon the people or draughts on the treasury, beyond the outlay for its first establishment.

As a first step towards the adoption of this system of telegraphs by the Government, your committee recommend the appropriation of thirty thousand dollars to be expended under the direction of the Postmaster General, in constructing a line of electro magnetic telegraphs, under the superintendence of Professor Sam'l F. B. Morse, of such length and between such points as shall fully test its practicability and utility; and for this purpose they respectfully submit the following bill:

A bill to test the Practicability of Establishing a System of Electro Magnetic Telegraphs by the United States.

Be it enacted by the Senate and House of Representatives of the United States in Congress assembled, That the sum of thirty thousand dollars be, and is hereby, appropriated, out of any moneys in the treasury not otherwise appropriated, for testing the capacity and usefulness of the system of electro magnetic telegraphs invented by Samuel F. B. Morse, of New York, for the use of the Government of the United States, by constructing a line of said electro magnetic telegraphs, under the superintendence of Professor Samuel F. B. Morse, of such length and between such points as shall fully test its practicability and utility; and that the same shall be expended under the direction of the Postmaster General, upon the application of said Morse.

SEC. 2. *And be it further enacted*, That the Postmaster General be, and he is hereby, authorized to pay, out of the aforesaid thirty thousand dollars, to the said Samuel F. B. Morse, and the persons employed under him, such sums of money as he may deem to be a fair compensation for the services of the said Samuel F. B. Morse and the persons employed under him, in constructing and in superintending the construction of the said line of telegraphs authorized by this bill.

No. 11.
Letter from Professor Henry to Professor Morse.

PRINCETON COLLEGE, *February 24, 1842.*

MY DEAR SIR: I am pleased to learn that you have again petitioned Congress in reference to your telegraph, and I most sincerely hope that you will succeed in convincing our representatives of the importance of the invention. In this you may, perhaps, find some difficulty, since, in the minds of many, the electro magnetic telegraph is associated with the various chimerical projects constantly presented to the public, and particularly with the schemes, so popular a year or two ago, for the application of electricity as moving power in the arts. I have asserted, from the first, that all attempts of this kind are premature, and made without a proper knowledge of scientific principles. The case is, however, entirely different in regard to the electro magnetic telegraph. *Science is now fully ripe for this application*, and I have not the least doubt, if proper means be afforded, of the perfect success of the invention.

The idea of transmitting intelligence to a distance by means of electrical action, has been suggested by various persons, from the time of Franklin to the present; but until within the last few years, or since the principal discoveries in electro magnetism, all attempts to reduce it to practice were necessarily unsuccessful. The mere suggestion, however, of a scheme of this kind is a matter for which little credit can be claimed, since it is one which would naturally arise in the mind of almost any person familiar with the phenomena of electricity; but the bringing it forward at the proper moment when the developments of science are able to furnish the means of certain success, and the devising a plan for carrying it into practical operation, are the grounds of a just claim to scientific reputation as well as to public patronage.

About the same time with yourself, Professor Wheatstone, of London, and Dr. Steinheil, of Germany, proposed plans of the electro magnetic telegraph, but these differ as much from yours as the nature of the common principle would well permit; and unless some essential improvements have lately been made in these European plans, I should prefer the one invented by yourself.

With my best wishes for your success, I remain, with much esteem yours, truly,

JOSEPH HENRY.

PROFESSOR MORSE.

No. 12.

Report of the American Institute on the Electro Magnetic Telegraph.

NEW YORK, *September 12, 1842.*

The undersigned, the committee of arts and sciences of the American Institute, respectfully report:

That, by virtue of the power of adding to their numbers, they called to their aid the gentlemen whose names are hereunto annexed, with those of the original members of the committee, and proceeded to examine Professor Morse's electro magnetic telegraph.

Having investigated the scientific principles on which it is founded, inspected the mechanism by which these principles are brought into practical operation, and seen the instruments in use in the transmission and return of various messages, they have come to the conclusion that it is admirably adapted to the purposes for which it is intended, being capable of forming words, numbers, and sentences, nearly as fast as they can be written in ordinary characters, and of transmitting them to great distances with a velocity equal to that of light. They, therefore, beg leave to recommend the telegraph of Professor Morse for such testimonials of the approbation of the American Institute as may in its judgment be due to a most important practical application of high science, brought into successful operation by the exercise of much mechanical skill and ingenuity.

All which is respectfully submitted.

JAMES RENWICK, LL. D.,
Prof. Chem. and Nat. Phil., Columbia Col., N. Y.

JOHN W. DRAPER, M. D.,
Prof. Chem. and Min., University, city of New York.

WILLIAM H. ELLET, M. D.
Prof. Chem., &c. Col. of Columbia, S. C.

JAMES R. CHILTON, M. D.,
Chem., &c., New York.

G. C. SCHAEFFER,
Associate Prof. Chem., Columbia Col., N. Y.

EDWARD CLARK.
CHARLES A. LEE, M. D.

Extract from the minutes of the Institute:

Resolved, That the report be accepted, adopted, and referred to the premium committee, and that the recording secretary be directed to publish the same, at the expense of the Institute.

No. 13.
Letter from S. F. B. Morse to the Hon. C. G. Ferris.

NEW YORK, *December 6, 1842.*

DEAR SIR: In compliance with your request, I give you a slight history of my electro magnetic telegraph, since it was presented for the consideration of Congress, in the year 1838.

During the session of the 25th Congress, a report was made by the Committee on Commerce of the house, which concluded by unanimously submitting a bill appropriating $30,000 for the purpose of testing my system of electro magnetic telegraphs. The pressure of business at the close of that session prevented any action being taken upon it.

Before the session closed, I visited England and France, for the double purpose of submitting my invention to the test of European criticism, and to secure to myself some remuneration for my large expenditures of time and money in elaborating my invention. In France, after a patent had been secured in that country, my telegraph first attracted the attention of the Academy of Sciences, and its operation was shown, and its principles were explained, by the celebrated philosopher, Arago, in the session of that distinguished body of learned men, on September 10, 1838. Its reception was of the most enthusiastic character. Several other societies, among which were the Academy of Industry and the Philotechnic Society, appointed committees to examine and report upon the invention, from all which I received votes of thanks, and from the former the large medal of honour. The French Government at this time had its attention drawn to the subject of electric telegraphs, several systems having been presented for its consideration, from England, Germany and France. Through the kind offices of our minister at the French Court, General Cass, my telegraph was also submitted; and the Minister of the Interior (M. Montalivet) appointed a commission, at the head of which was placed M. Alphonse Foy, the administrator-in-chief of the telegraphs of France, with directions to examine and report upon all the various systems which had been presented. The result of this examination (in which the ingenious systems of Professor Wheatstone, of London, of Professor Steinheil, of Munich, and Professor Masson, of Caen, passed in review) was a report to the Minister in favor of mine. In a note addressed to me by M. Foy, who had expressed his warmest admiration of my telegraph in my presence, he thus writes:

"I take a true pleasure in confirming to you in writing that which I have already had the honor to say to you viva voce, that I have prominently presented (signalé) to Monsieur the Minister of the Interior your electro magnetic telegraph, as being the system which presents the best chance of a practical application; and I have stated to him that if some trials are to be made with electric telegraphs, I hesitate not to recommend that they should be made with your apparatus."

In England, my application for a patent for my invention was opposed before the Attorney General by Professor Wheatstone and Mr. Davy, each of whom had systems already patented, essentially like each other, but very different from mine. A patent was denied me by the Attorney General, Sir John Campbell, on a plea which I am confident will not bear a legal examination. But there being no appeal from the Attorney General's decision, nor remedy, except at enormous expense, I am deprived of all benefit from my invention in England. Other causes than impartial justice evidently operated against me. An interest for my invention, however, sprung up voluntarily, and quite unexpectedly, among the English nobility and gentry in Paris, and, had I possessed the requisite funds to prosecute my rights before the British Parliament, I could scarcely have failed to secure them, so powerfully was I supported by this interest in my favour; and I should be ungrateful did I not take every opportunity to acknowledge the kindness of the several noblemen and gentlemen who volunteered to aid me in obtaining my rights in England, among the foremost of whom were the Earl of Lincoln, the late celebrated Earl of Elgin, and the Hon. Henry Drummond.

I returned to the United States in the spring of 1839, under an engagement entered into in Paris with the Russian Counsellor of State, the Baron Alexandre de Meyendorff, to visit St. Petersburg with a distinguished French savant, M. Amyot, for the purpose of establishing my telegraphic system in that country. The contract, formally entered into, was transmitted to St. Petersburg, for the signature of the Emperor, which I was led to believe would be given without a doubt; and, that no time should be lost in my preparations, the contract, duly signed, was to be transmitted to me in in New York, through the Russian ambassador in the United States, in four or five weeks, at farthest, after my arrival home.

After waiting, in anxious suspense, for as many months, without any intelligence, I learned *indirectly* that the Emperor, from causes not satisfactorily explained, refused to sign the contract.

These disappointments, (not at all affecting the scientific or practical character of my invention,) combined with the financial depression of the country, compelled me to rest a while from further prosecuting my enterprise. For the last two years, however, under many discouraging circumstances, from want of the requisite funds for more thoroughly investigating some of the principles involved in the invention, I have, nevertheless, been able to resolve all the doubts that lingered in my own mind, in regard to the perfect practicability of establishing my telegraphic system to any extent on the globe. I say, "doubts that lingered in my own mind;" the principle, and, indeed, the only one of a scientific character, which at all troubled me, I will state, and the manner in which it has been resolved:

At an early stage of my experiments, I found that the magnetic power produced in an electro magnet, by a single galvanic pair, diminished rapidly as the length of the conductors increased. Ordinary reasoning on this fact would lead to a conclusion fatal to the whole invention, since at a great distance I could not operate at all, or, in order to operate, I should be compelled to make use of a battery of such a size as would render the whole plan in effect impracticable. I was, indeed, aware, that by multiplying the pairs in the battery—that is, increasing the intensity of its propulsive power, certain effects could be produced at great distances, such as the decomposition of water, a visible spark, and the deflection of the magnetic needle. But as magnetic effects, except in the latter case, had not, to my knowledge been made the subject of careful experiment, and as these various effects of electrical action seemed, in some respects, to be obedient to different laws, I did not feel entirely assured that magnetism could be produced by a multiplication of pairs sufficiently powerful at a great distance to effect my purpose. From a series of experiments which I made, in conjunction with Professor Fisher, during the last summer, upon 33 miles of wire, the interesting fact so favorable to my telegraphic system, was fully verified, that *while the distance increased in an arithmetical ratio, an addition to the series of galvanic pairs of plates increased the magnetic power in a geometric ratio.* Fifty pairs of plates were used as a constant power. Two miles of conductors at a time, from two to thirty-three, were successively added to the distance. The weight upheld by the magnet from the magnetism produced by 50 pairs, gradually diminished up to the distance of 10 miles; after which, *the addition of miles of wire up to 33 miles* (the extent to which we were able to try it) *caused no further visible diminution of power.* The weight then sustained was a constant quantity. The practical deduction from these experiments is the fact that with a very small battery all the effects I desire, and at any distance, can be produced. In the experiments alluded to, the fifty pairs did not occupy a space of more than 8 cubic inches, and they comprised but 50 square inches of active surface.

The practicability of establishing my telegraphic system is thus relieved from all scientific objections.

Let me now turn your attention, sir, one moment to a consideration of the telegraph as a source of revenue. The imperfections of the common systems, particularly their uselessness, on account of the weather, three quarters of the time, have concealed from view so natural a fruit of a perfected telegraphic system. So uncertain are the common telegraphs as to time, and so meager in the quantity of intelligence they can transmit under the most favorable circumstances, that the idea of making them a source of revenue would not be likely to occur. So far, indeed, from being a source of revenue, the systems in common use in Europe are sustained at great expense; an expense which, imperfect as they are, is justified, in the view of the Government, by the great political advantages which they produce. Telegraphs with them are a Government monopoly, and used only for Government purposes. They are in harmony with the genius of those Governments. The people have no advantage from them, except indirectly as the Government is benefitted. Were our mails used solely for the purpose of the Government, and private individuals forbidden to correspond by them, they would furnish a good illustration of the operation of the common European telegraphic systems.

The electro magnetic telegraph, I would fain think, is more in consonance with the political institutions under which we live, and is fitted, like the mail system, to diffuse its benefits alike to the Government and to the people at large.

As a source of *revenue*, then, to the Government, few, I believe, have seriously computed the great profits to be derived from such a system of telegraphs as I propose; and yet there are sure data already obtained by which they can be demonstrated.

The first fact is, that every minute of the 24 hours is available to send intelligence.

The second fact is, that 12 signs, at least, can be sent in a minute, instantaneously, as any one may have proof by actual demonstration of the fact on the instrument now operating in the capitol.[15]

There can be no doubt that the cases, where such speedy transmission of intelligence from one distant city to another is desirable, are so numerous, that when once the line is made for such transmission, it will be in constant use, and a demand made for a greater number of lines.

The paramount convenience, to commercial agents and others, of thus corresponding at a distance, will authorize *a rate of postage proportionate to the distance*, on the principle of rating postage by the mails.

To illustrate the operation of the telegraph in increasing the revenue, let us suppose that but 18 hours of the 24 are efficiently used for the actual purposes of revenue; that 6 hours are allowed for repetitions and other purposes, which is a large allowance. This would give, upon a single circuit, 12,960 signs per day, upon which a rate of postage is to be charged. Intelligence of great extent may be comprised in a few signs. Suppose the following commercial communication is to be transmitted from New York to New Orleans:

Yrs., Dec. 21, rec. Buy 25 bales c., at 9, and 300 pork, at 8.

Here are 36 signs, which take three minutes in the transmission from New York to New Orleans, and which informs the New York merchant's correspondent at New Orleans of the receipt of a certain document, and gives him orders to purchase 25 bales of cotton at 9 cents per pound, and 300 barrels of pork at 8 cents per pound. Thus may be completed, in three minutes, a transaction in business which now would take at least four or five weeks to accomplish.

Suppose that one cent per sign be charged for the first 100 miles, increasing the charge at the rate of half a cent each additional 100 miles, the postage of the above communication would be $2.88 for a distance of 1,500 miles. It would be sent 100 miles for 36 cents. Would any merchant grudge so small a sum for sending such an amount of information in so short a time to such a distance? If time is money, and to save time is to save money, surely such an immense saving of time is the saving of an immense sum of money. A telegraphic line of a single circuit only, from New York to New Orleans, would realize, then, to the Government, *daily*, in the correspondence between those two cities alone, over *one thousand dollars* gross receipts, or over $300,000 per annum.

But it is a well-established fact, that, as facilities of intercourse increase between different parts of the country, the greater is that intercourse. Thousands travel, in this day of rail roads and steamboats, who never thought of leaving their homes before. Establish, then, the means of instantaneous communication between the most distant places, and the telegraphic line of a single circuit will very soon be insufficient to supply the demands of the public—they will require more.

Two circuits will of course *double the facilities, and double the revenue*; but it is an important fact, that the expense of afterwards establishing a second, or any number of circuits, does not proceed on the *doubling* principle. If a channel for conveying a single circuit be made, in the first instance, of sufficient capacity to contain many more circuits, which can easily be done, additional circuits can be laid as fast as they are called for, at but little more than the cost of the prepared wire. The recent discovery of Professor Fisher and myself, shows that a single wire may be made the common conductor for at least six circuits. How many more we have not yet ascertained. So that, to add another circuit is but to add another wire. Fifty dollars per mile under these circumstances, would therefore add the means of doubling the facilities and the revenue.

Between New York and Philadelphia, for example, the whole cost of laying such an additional circuit would be but $5,000, which would be more than defrayed by *two months'* receipts only from the telegraphs between those two cities.

There are two modes of establishing the line of conductors.

The first and cheapest is doubtless that of erecting spars about 30 feet in height and 350 feet apart, extending the conductors along the tops of the spars. This method has some obvious disadvantages. The expense would be from $350 to 400 per mile.

The second method is that of enclosing the conductors in leaden tubes, and laying them in the earth. I have made the following estimate of the cost of this method:

Wire, prepared, per mile,	$ 150.00
Lead pipe, with solderings,	250.00
Delivery of the pipe and wire,	25.00
Passing wire into the pipes,	5.00
Excavations and filling in about 1,000 yards per mile,	
or 3 feet deep, at 15 cents per square yard,	150.00
Laying down the pipe,	3.00
	583.00
One register, with its machinery, comprising a galvanic	
battery of four pairs of my double-cup battery,	$ 100.00
One battery of 200 pairs,	100.00
Expense for thirty nine miles,	$ 22,837.00
Two registers,	200.00
Two batteries,	200.00
Services of chief superintendent of construction, per annum,	2,000.00
Services of three assistants, at $1,500 each, per annum,	4,500.00
	$ 29,637.00

As experience alone can determine the best mode of securing the conductors, I should wish the means and opportunity of trying various modes, to such an extent as will demonstrate the best.

Before closing my letter, sir, I ought to give you the proofs I possess that the American telegraph has the *priority in the time of its invention.*

The two European telegraphs in practical operation are Professor Steinheil's of Munich, and Professor Wheatstone's of London. The former is adopted by the Bavarian Government; the latter is established about 200 miles in England, under the direction of a company in London. In a highly interesting paper on the subject of telegraphs, translated and inserted in the London Annals of Electricity, March and April, 1839, Professor Steinheil gives a brief sketch of all the various projects of electric telegraphs, from the time of Franklin's electrical experiments to the present day. Until the birth of the science of electro magnetism, generated by the important discovery of Oersted, in 1820, of the action of electric currents upon the magnetic needle, the electric telegraph was but a philosophic toy, complicated and practically useless. Let it be here noticed, that, after this discovery of Oersted, the *deflection of the needle* became the principle upon which the savants of Europe based all their attempts to construct an electric telegraph. The celebrated Ampère, in the same year of Oersted's discovery, suggested a plan of telegraphs, to consist of a magnetic needle, and a circuit for each letter of the alphabet and the numerals—making it necessary to have some 60 or 70 wires between the two termini of the telegraphic line.

This suggestion of Ampère is doubtless the parent of all the attempts in Europe, both abortive and successful, for constructing an electric telegraph.

Under this head may be arranged the Baron Schilling's at St. Petersburg, consisting of 36 magnetic needles, and upwards of 60 metallic conductors, and invented, it seems, at the same date with my electro magnetic telegraph, in the autumn of 1832. Under the same head comes that of professors Gauss and Weber, of Göttingen, in 1833, who simplified the plan by using but a single needle and a single circuit. Professor Wheatstone's of London, invented in 1837, comes under the same category; he employs five needles and six conductors. Professor Steinheil's, also invented in 1837, employs two needles and two conductors.

But there was another discovery, in the infancy of the science of electro magnetism, by Ampère and Arago, immediately consequent on that of Oersted, namely: the *electro magnet*, which none of the savants of Europe who have planned electric telegraphs ever thought of applying, until within two years past, for the purpose of signals. My telegraph is essentially based on this latter discovery.

Supposing my telegraph to be based on the same principle with the European electric telegraphs, which it is not, mine, having been invented in 1832, would still have the precedence, by some months at least, of Gauss and Weber's, to whom Steinheil gives the credit of being the first to simplify and make practicable the electric telegraph. But when it is considered that all the European telegraphs make use of the deflection of the needle to accomplish their results, and that none use the *attractive power of the electro magnet to write in legible characters*, I think I can claim, without injustice to others, to be the first inventor of the *electro magnet telegraph.*

In 1839, I visited London, on my return from France, and through the polite solicitations of the Earl of Lincoln, showed and explained its operation at his house, on the 19th of March, 1839, to a large company which he had expressly invited for the purpose, composed of Lords of the Admiralty, members of the Royal Society, and members of both Houses of Parliament.

Professor Wheatstone has announced that he has recently (in 1840) also invented and patented an *electro magnetic telegraph*, differing altogether from his invention of 1837, which he calls his *magnetic needle telegraph.* His is, therefore, the first European electro magnetic telegraph, and was invented, as is perceived, eight years subsequent to mine, and one year after my telegraph *was exhibited in the public manner described at the Earl of Lincoln's residence in London.*

I am the more minute in adducing this evidence of priority of invention to you, sir, since I have frequently been charged by Europeans, in my own country, with merely imitating long known European inventions. It is, therefore due to my own country, as well as to myself, that in this matter the facts should be known.

Professor Steinheil's telegraph is the only European telegraph that professes to *write* the intelligence. He records, however, by the delicate touch of the needle in its deflections, with what practical effect I am unable to say; but I should think that it was too delicate and uncertain, especially as compared with the strong and efficient power which may be produced in any degree by the electro magnet.

I have devoted many years of my life to this invention, sustained in many disappointments by the belief that it is destined eventually to confer immense benefits upon my country and the world.

I am persuaded that whatever facilitates intercourse between the different portions of the human family will have the effect, under guidance of sound moral principles, to promote the best interests of man. I ask of Congress the means of demonstrating its efficiency.

I remain, sir, with great respect, your most obedient servant,

SAM. F. B. MORSE.

Hon. CHARLES G. FERRIS,

Member of the House of Representatives from the city of New York and one of the Committee on Commerce, to whom was referred the subject of the expediency of adopting a system of electro magnetic telegraphs for the United States.

No. 14.

Communication from the Secretary of the Treasury, transmitting the Report of Professor Morse, announcing the completion of the Electro Magnetic Telegraph between the cities of Washington and Baltimore. June 6, 1844. Referred to the Committee on Commerce.

TREASURY DEPARTMENT, *June 4, 1844.*

SIR: I have the honour respectfully to transmit herewith, for information of the House of Representatives, a report, dated the 3d instant, from Professor Sam. F. B. Morse, announcing the completion of the electro magnetic telegraph between Washington and the city of Baltimore, as authorized by the "Act to test the practicability of establishing a system of electro magnetic telegraphs by the United States," approved the 3d of March, 1843.

I beg leave to state, that the perfect practicability of the system has been fully and satisfactorily established by the work already completed.

The subject is respectfully submitted to the consideration of Congress for such further directions in the matter as may be deemed expedient.

I have the honor to be, very respectfully, your obedient servant,

McCLINTOCK YOUNG,
Secretary of the Treasury ad interim.

Hon. JOHN W. JONES,
Speaker of the House of Representatives.

No. 15.

Letter from Professor Morse to Hon. McClintock Young.

WASHINGTON, *June 3, 1844.*

SIR: I have the honour to report that the experimental essay authorized by the act of Congress on March 3d, 1843, appropriating $30,000 for "testing" my "system of electro magnetic telegraphs, and of such length, and between such points, as shall test its *practicability* and *utility*," has been made between Washington and Baltimore—a distance of forty miles—connecting the capitol in the former city, with the rail road depot in Pratt street, in the latter city.

On the first point proposed to be settled by the experiment—to wit, its *practicability*—it is scarcely necessary to say (since the public demonstration which has been given of its efficacy, for some days past, during the session of the different conventions in the city of Baltimore) that it is fully proved.

Items of intelligence of all kinds have been transmitted back and forth, from the simple sending of names, to the more lengthened details of the proceedings of Congress and the conventions. One fact will, perhaps, be sufficient to illustrate the efficiency and speed with which intelligence can be communicated by the telegraph.

In the proceedings of the democratic convention at Baltimore for the nomination of a candidate for President of the United States at the next election, the result of the votes in the nomination of the Hon. J. K. Polk was conveyed from the convention to the telegraphic terminus in Baltimore, transmitted to Washington, announced to the hundreds assembled in front of the terminus at the Capitol, and to both Houses of Congress; the reception of the news at Washington was then transmitted to Baltimore, sent to the convention and circulated among its members—all before the nomination of the successful candidate was *officially announced* by the presiding officer of the convention.

In regard to the *utility* of the telegraph, time alone can determine and develop the whole capacity for good of so perfect a system. In the few days of its infancy, it has already casually shown its usefulness in the relief, in various ways, of the anxieties of thousands; and, when such a sure means of relief is available to the public at large, the amount of its usefulness becomes incalculable.

An instance or two will best illustrate this quality of the telegraph:

A family in Washington was thrown into great distress by a rumor that one of its members had met with a violent death in Baltimore the evening before. Several hours must have elapsed ere their state of suspense could be relieved by the ordinary means of conveyance. A note was despatched to the telegraph rooms at the Capitol, requesting to have inquiry made at Baltimore. The messenger had occasion to wait but *ten minutes*, when the proper inquiry was made at Baltimore, and the answer returned that the rumor was without foundation. Thus was a worthy family relieved immediately from a state of distressing suspense.

An inquiry from a person in Baltimore holding the check of a gentleman in Washington, upon the Bank of Washington, was sent by telegraph, to ascertain if the gentleman in question had funds in that bank. A messenger was instantly despatched from the Capitol, who returned in a few minutes with an affirmative answer, which was returned to Baltimore instantly; thus establishing a confidence in a money arrangement, which might have affected unfavorably (for many hours at least) the business transactions of a man in good credit.

Other cases might be given; but these are deemed sufficient to illustrate the point of utility, and to suggest to those who will reflect upon them, thousands of cases in the public business, in commercial operations, and in private and social transactions, which establish beyond a doubt the immense advantages of such a speedy mode of conveying intelligence.

In the construction of this *first line of conductors*, it was necessary that experiments should be made to ascertain the best mode of establishing them. The plan I first suggested in my letter to the Secretary of the Treasury in 1837, (see the House report, No. 6, April 6, 1838,) of placing my conductors upon posts thirty feet high, and some three hundred feet apart, is, after experiment, proved to be the most eligible. The objection, so strongly urged in the outset, that, by being exposed above ground, the conductors were in danger from evil disposed persons, had such weight with me, in the absence of experience on the subject, as early to turn my whole attention to the practicability of placing my conductors in tubes beneath the earth, as the best means of safety. The adoption of this latter mode, for some thirteen miles in England, by the projectors of the English telegraph, confirmed me in the belief that this would be best. I was thus led to contract for lead pipe sufficient to contain my conductors through the whole route. Experience, however, has shown that this mode is attended with disadvantages far outweighing any advantages from its fancied security beneath the ground. If apparently more secure, an injury once sustained is much more difficult of access, and of repair; while upon posts, if injury is sustained, it is at once seen, and can be repaired, ordinarily almost without cost. But the great advantage of the mode on posts over that beneath the ground, is the cheapness of its construction. This will be manifest from the following comparative estimate of the two modes in England and in America:

Cost of English Telegraph.

In pipe, £287 6*s*., or $1,275 per mile.

On posts, £149 5*s*., or $662 per mile.

Cost of American Telegraph, as estimated in House Report, No. 17, 27th Congress, 3d session.

In pipe, $583 per mile.

On posts, from $350 to $400 per mile.

These comparisons also show how much less is the cost of the American telegraph, even at the highest estimate.

But these estimates of the cost of construction, largely exceed the actual cost, under the improved modes recently suggested by experiment, and now adopted; and the cost of the line between Baltimore and Washington, already constructed, involves numerous expenditures of an experimental character, which will not be incident to an extension of the line onward to New York, if that shall be deemed desirable.

Of the appropriation made, there will remain in the treasury, after the settlement of outstanding accounts, about $3,500, which may be needed for contingent liabilities, and for sustaining the line already constructed, until provision by law shall be made for such an organization of a telegraphic department or bureau as shall enable the telegraph at least to support itself, if not to become a profitable source of revenue to the Government.

I will conclude by saying, that I feel grateful for the generous confidence which Congress has thus far extended toward me and my enterprise; and I will cheerfully afford any further and more detailed information on the subject of the telegraph, when desired, and will be prepared to make and execute any desirable arrangements for the extension of it that Congress shall require.

With great respect, your obedient servant,

SAM. F. B. MORSE,
Superintendent of Electro Magnetic Telegraph.

To the Hon. MCCLINTOCK YOUNG,
Secretary of the Treasury ad interim.

No. 16.

Letter from the Secretary of the Treasury, transmitting a letter from Professor Morse, relative to the Magnetic Telegraph, Dec'r 23, 1844.

TREASURY DEPARTMENT, *December 17, 1844.*

SIR: In compliance with the request made in your letter of this date, in behalf of the Committee on Commerce of the House of Representatives, for information from this department upon the subject of "Morse's telegraph," I have the honour respectfully to transmit herewith a communication from Professor Morse, dated the 12th instant, containing specific information in regard to that work.

I have the honour to be, very respectfully, your obedient servant,

GEO. M. BIBB,
Secretary of the Treasury.

Hon. ISAAC E. HOLMES,
Chairman of Committee on Commerce,
House of Representatives.

No. 17.
Letter from Prof. Morse to the Hon. G. M. Bibb.

WASHINGTON, *December 12, 1844.*

SIR: I have the honour respectfully to submit some facts in relation to the electro magnetic telegraph, bearing upon the bill now before Congress, reported from the Committee on Commerce of the House, for the extension of the telegraphic line from Baltimore to New York.

By a reference to documents in the records of the government, it will appear, that the subject of establishing a system of telegraphs for the use of the United States has been, occasionally, for many years, before Congress; but nothing effective was ever done in relation to the matter, until the Hon. Levi Woodbury, while Secretary of the Treasury, by addressing circular letters to various individuals in the United States, (among which was one to me,) drew forth from me a general description of the advantages of a system of electro magnetic telegraphs which I had invented in 1832, on my passage from France to the United States. For my answer to this circular letter, see No. 2, taken from House report, No. 753, 25th Congress, second session; and I refer to it now, to show that the assertions respecting the practicability and utility of my system have been fully and satisfactorily sustained by the result of the experimental essay, authorized by the government, establishing the line between Washington and Baltimore.

That which seemed to many chimerical at the time, is now completely realized. The most sceptical are convinced; and the daily and hourly operations of the telegraph in transmitting information of any kind are so publicly known, and the public feeling in regard to it so universally expressed, that I need here only give a few instances of its action, further to illustrate its character.

The facts in relation to the transmission of the proceedings of the democratic convention of Baltimore, in May last are well known, and are alluded to in my report to the department, June 3d, 1844, No. 15. Since the adjournment of Congress in June last, and during the summer and the autumn, the telegraph has been in constant readiness for operation, and there has been time to test many points in relation to it, which needed experience to settle.

For more now than *eight months*, the conductors for the telegraph, carried on elevated posts for 40 miles, have remained undisturbed from the wantonness or evil disposition of any one. Not a single instance of the kind has occurred. In several instances, indeed, the communication has been interrupted by accidents, but then only for a very brief period. One of these was by the great fire in Pratt street, Baltimore, which destroyed one of the posts, and consequently, temporarily stopped the communication; but in two or three hours the damage was repaired, and the first notice of the accident and all the particulars were transmitted to Washington by the telegraph itself.

Another instance of interruption was occasioned by the falling of a tree, which accidently fell across the wires, and at the same time across the rail road track, stopping the cars for a short time, and the telegraphic communication for two hours.

Excepting the time excluded by these, and two or three other similar accidental interruptions, and which, during seven months of its effective existence between the two cities, does not altogether amount to more than 24 hours, the telegraph has been either in operation, or prepared for operation, at any hour of the day or night, irrespective of the state of the weather.

It has transmitted intelligence of great importance. During the troubles in Philadelphia the last summer, sealed despatches were sent by express from the Mayor of Philadelphia to the President of the United States. On the arrival of the express at Baltimore, the purport of the despatches transpired; and while the express train was in preparation for Washington, the intelligence was sent to Washington by telegraph, accompanied by an order from the president of the rail road company to prevent the Washington burden train from leaving until the express should arrive. The order was given and complied with. The express had a clear track, and the President and the Cabinet (being in council) had notice both of the fact that an express was on its way with important despatches to them, and also of the nature of those despatches, so that, when the express arrived, the answer was in readiness for the messenger.

In October, a deserter from the U.S. ship Pennsylvania, lying at Norfolk, who had defrauded also the purser of the ship of some $600 or $700, was supposed to have gone to Baltimore. The purser called at the telegraph office in Washington, stated his case, and wished to give notice in Baltimore, at the same time offering a reward for the apprehension of the culprit. The name and description of the offender's person, with the offer of the reward, were sent to Baltimore, and in ten minutes the warrant was in the hands of the officers of justice for his arrest; and in half an hour from the time that the purser profferred his request at Washington, it was announced from Baltimore by the telegraph, "The deserter is arrested; he is in jail; what shall be done with him?"

To show the variety of the operations of the telegraph, a game of draughts, and several games of chess, have been played between the cities of Baltimore and Washington, with the same ease as if the players were seated at the same table. To illustrate the independence of the telegraph of the weather, and time of day, I would state that, during the severe storm of the 5th December, when the night was intensely dark, the rain descending in torrents, and the wind blowing a gale, it seemed more than ordinarily mysterious to see a company around a table, in a warm retired chamber, on such a night, in Washington, playing a game of chess with another company similarly situated in Baltimore: the darkness, the rain, and the wind, being no impediment to instantaneous communication.

In regard to the quantity of intelligence which may be sent in a given time, it is perfectly safe to say that thirty characters can be transmitted in a minute by a single instrument; and as these characters are conventional signs, they may mean either *numbers*, *letters*, *words*, or *sentences*. As an illustration of this point, I will state that nearly a whole column (more than seven-eighths) in the Baltimore Patriot was transmitted in thirty minutes—faster than the reporter in Baltimore could transcribe.

This fact bears upon the ability of producing a revenue from the telegraph; and I would suggest the propriety of permission being granted by Congress to the department, to adjust a tariff of charges on intelligence sent by telegraph, at such a rate of postage as shall at least return to the treasury the interest of the capital expended in the first construction, and after maintenance of the telegraph.

In aid of this view of the subject, I beg to refer to my letter to the chairman of the Committee on Commerce, December 6, 1842, No. 13.

Since that was written, experience has shown that that calculation is far below the real results. Instead of *twelve signs* in a minute, upon which that computation was based, we must substitute *thirty*—a column of a newspaper having been transmitted to Baltimore even at the rate of *thirty-five* signs in a minute. It is, therefore, safe to set down the rate at 30 signs per minute; and it is safe to double the annual receipts, making the gross amount $600,000 per annum.

In the absence of experience, the expense necessary to construct and to maintain a system of electro magnetic telegraphs, was thought to be so great as to present a formidable, if not an insurmountable obstacle to its adoption.

But the experiment already made for 40 miles, has shown that the electro magnetic telegraph is far from being expensive, either in its first construction, or after maintenance, especially when its vast superiority over the old system is taken into consideration.

To make this more clear, I give an abstract both of the expenses and capacities of the ordinary visual telegraphs in some of the European countries.

In England, the semaphore telegraph, established between London and Portsmouth, a distance of 72 miles, is maintained by the British government at an average expense of £3,405, or $15,118 per annum. From a return [vol. 30, 1843, accounts and papers of House of Commons] of the number of days during which the telegraph was *not available*, on account of the weather, during a period of three years, it appears that there were, in that time, 323 days in which it was useless, or nearly *one year out of three*! But by a return made to the admiralty of the number of hours in the day appointed for working the telegraph, it appears that the hours appointed for the year are—from 1st October to 28th February, from 10 o'clock, a. m., to 3 p. m.; 5 hours. From 1st March to 30th September, from 10 a. m., to 5 p. m.; 7 hours.

Average number of hours per day, in the most favourable weather, 6 hours!

Deducting 1 year from the 3, for unavailable days, the average time per day for the 3 years would be but 4 hours. So that, for the use of their telegraph for 72 miles, and for only 4 hours in the day, the British government expend $15,118 per annum.

The French system of telegraphs is more extensive and perfect than that of any other nation. It consists, at present, of five great lines, extending from the capital to the extreme cities of the kingdom, to wit:

The Calais line, from Paris to Calais,	152	miles
The Strasbourg line, from Paris to Strasbourg,	255	"
The Brest line, from Paris to Brest,	325	"
The Toulon line, from Paris to Toulon,	317	"
The Bayonne line, from Paris to Bayonne,	425	"
	1,474	miles.

Making a total of 1,474 miles of telegraphic intercourse. These telegraphs are maintained by the French government at an annual expense of over 1,000,000 of francs, or $202,000.

The whole extent, then, of the French lines of telegraph is 1,474 miles, with 519 stations; and (if the estimate for six stations, at an average cost of 4,400 francs, is a criterion for the rest) erected at a cost of at least $880 each—making a total of $456,720.

The electro magnetic telegraph, at the rate proposed in the bill, to wit, $461 per mile, (and which, it should be remembered, will construct not *one* line, only, but *six*,) could be constructed the same distance for $619,514—not one-third more than the cost of the French telegraphs. Even supposing each line to be only as efficient as the French telegraph, still there would be six times the facilities, for not one-third more cost. But when it is considered that the French telegraph, like the English, is unavailable the greater part of the time, the advantages in favour of the magnetic telegraph become more obvious.

An important difference between the two systems is, that the foreign telegraphs are all a burden upon the treasury of their respective countries; while the magnetic telegraph proposes, and is alone capable of sustaining itself and of producing a revenue.

Another difference in the two systems is, that the stations in the foreign telegraphs must be within sight of each other: a fact which bears essentially on the cost of maintenance. The French telegraph requires for the distance of 1,474 miles, no less than 519 stations—averaging *one for about every three miles.* The number of stations of the magnetic telegraph, on the contrary, is optional. The two stations (one only at Baltimore, and one at Washington) show that they may be at least 40 miles apart; and there is no reason to doubt, from experiments I have made, that 100 miles, or even 500 miles, would give the same results. In the maintenance, therefore, of stations, the magnetic telegraph would require but 15 stations, (assuming that 100 miles is the *utmost limit* of transmission between two stations which is not probable;) while the French requires 519 for the same distance.

When to this are added the facts that the magnetic telegraph is at *all times available*, at *every hour of the day or night, irrespective of weather*; that, in comparison with the visual telegraphs, it communicates *more than a hundred-fold* the quantity of intelligence in the same time; that it is originally constructed at a *less cost*, (*all things considered;*) that it is *maintained for less*; and that it is capable, by a rate of charges for transmitting intelligence, not only of defraying all its expenses, but, if desired, of producing a revenue, I may be permitted to hope that when these great advantages are fully understood, my system will receive that attention from the government which its intrinsic public importance demands.

I have as yet said nothing on the telegraph as a mighty aid to national defence. Its importance in this respect is so obvious, that I need not dilate.

The importance generally to the government and to the country, of a *perfect* telegraphic system, can scarcely be estimated by the short distance already established between Baltimore and Washington. But when all that transpires of public interest at New Orleans, at St. Louis, at Pittsburgh, at Cincinnati, at Buffalo, at Utica, at Albany, at Portland, at Portsmouth, at Boston, at New York, at Philadelphia, at Baltimore, at Washington, at Norfolk, at Richmond, at Charleston, at Savannah, and at all desired intermediate points, shall be *simultaneously* known in each and all these places together—when all the agents of the government, in every part of the country, are in instantaneous communication with head-quarters—when the several departments can at once learn the actual existing condition of their remotest agencies, and transmit at the moment their necessary orders to meet any exigency—then will some estimates be formed both of the powers and advantages of the magnetic telegraph.

Should the government be now disposed to possess the right of the proprietors, by giving them a fair consideration, I shall be ready to treat with them on the terms of transfer.

For myself, I should prefer that the government should possess the invention, although the pecuniary interests of the proprietors induce them to lean towards arrangements with private companies.

In closing this report, I would take the opportunity of favorably mentioning to the department the efficient attention to the duties of their respective stations given by my assistants, Alfred Vail and H. J. Rogers, esqrs.—the former directing the correspondence at the Washington terminus, and the latter at the Baltimore terminus.

Very respectfully, sir, your obedient servant,

SAM. F. B. MORSE,
Superintendent of Electro Magnetic Telegraphs
for the United States.

To the Hon. GEO. M. BIBB,
Secretary of the Treasury.

Magnetic Telegraph from Baltimore to New York, March 3, 1845.

Mr. CHAPPELL, from the Committee of Ways and Means, made the following Report.

The Committee of Ways and Means, to whom was referred a resolution instructing said committee to inquire into the expediency of reporting a bill to continue the Electro Magnetic Telegraph from Baltimore to New York, by way of Philadelphia, beg leave to submit the following report:

The authority given by the constitution to Congress to establish post offices and post roads, so far as it operates to confer on the government any power which would not equally belong to it without that provision, amounts simply to making the government, a public or a common carrier of the written correspondence of individuals, and of the lighter form of printed intelligence and news. In other words, by virtue of this clause, the government is authorized and required to pursue, on a scale commensurate with the wants and extent of the country, the business of receiving, transporting, and delivering letters, newspapers, and pamphlets, for all persons, private, as well as public, and to and from any and all places in the Union. And for the service thus rendered, the government exacts from the individuals served, a specific fee or compensation, under the name of postage, for every letter or paper transported and delivered. Now, it is quite obvious that both the pursuit of this business, and the exaction of a remuneration for it, would be altogether beyond the range of federal authority, but for the specially granted power to establish post offices and post roads. Mere silence in the constitution on this subject would have effectually withheld the power from the general government, and would have caused the business of carrying letters, newspapers, &c., to remain where all other branches of the carrying trade are actually left—namely, in the hands of individual enterprise, subject to State legislation, and to such (and no other) federal control as is involved in the power of Congress to regulate commerce among the States.

The functions thus devolved on the government, of performing for the people the office of universal letter carrier and news carrier, is a matter of the very highest consequence in every light in which it can be viewed. The bare fact that our ancestors refused to leave it dependent on individual enterprise or State control, and vested it expressly in Congress, abundantly attested their anxious sense of its importance, and their conviction of the impracticability of realizing the requisite public advantages from it, otherwise than by giving it a federal lodgment and administration.

Had not these advantages been regarded as attainable in no other way, while, at the same time, they were felt to be virtually necessary, the framers and adopters of the constitution, devoted as they are known to have been to the power and importance of the States, and jealously apprehensive of the undue preponderance of the federal branch, would never have consented to engraft on that branch a power so great, so growing, so penetrating and pervading, as that of the post office system—a power involving the direct exercise of the carrying trade by the government on a vast scale, and requiring, in order to its exercise, the organization and maintenance of a huge and distinct administrative department, which, in its operations, touches daily and intimately the private affairs as well as public interests of the people; receives and expends millions of money every year; and continually employs, pays, and controls many thousands of persons, scattered through all parts of the country—thus adding mightily to federal power, and especially to the influence and patronage of the federal executive. These are all consequences which result directly and necessarily from the bestowment of the post office power on the general government. And inasmuch as the government thus derives from that power so great an addition to its own weight and influence, it certainly ought to be considered as contracting therefrom a correspondently heavy obligation to make the power advantageous and useful to the people, to the utmost extent of which it is capable.

The government has ever shown itself fully sensible of this obligation, and alive to its fulfilment. Hence, that immense and minute machinery of post offices and post roads, of postmasters, contractors, and carriers, which overspreads the country, and meets us everywhere—all designed and kept up for the sole purpose of bringing the contents of the mail-bag, with frequency, regularity, and celerity, near to the doors of our whole population. For many years, no better or more expeditious means of conveyance could be found than horse-power in the various forms in which it might be applied on ordinary highways. But in those times, as well as now, the government acted on the principle of not regarding even a heavy increase of expense as an objection sufficient to outweigh so important an object as the regular, frequent, and rapid transmission of the mail between all the great points, and along all the chief arteries of the country. On such routes, accordingly, the mail was kept running without interruption—by night as well as by day—and at the best speed that could be secured by a well organized and costly system of relays of men, horses, and vehicles.

But, at length, the ever advancing discoveries and improvements of science and art threw into the shade, as slow and inadequate, all the old and long used modes of travel and transportation. Steamboats and railroads burst upon the world, introducing a new and wonderful era in its commerce and intercourse; private capital and enterprise soon built them up, and put them in operation, whenever a sufficiently tempting prospect of gain appeared; and all private persons, as well as public departments, saw presented to their option more perfect and expeditious modes of transportation than could have possibly entered into the anticipations of the framers of the constitution. But, though not anticipated or foreseen, these new and improved modes were as clearly within the purview of the constitution, as were the older and less perfect ones with which our ancestors were familiar. And there being no doubt entertained either on this point, or as to the obligation of the government to lay hold of the best and most rapid methods of transmission which the improvements of the age put in its reach, steam-power commended itself at once to adoption, and has long been extensively employed, both on land and water, for the carriage of the mail.

It is not without full reflection that the committee insist on the principle that it was the duty as well as the right of the government thus to avail itself, even at heavy additional expense, of the powerful agency of steam, for the purpose of accelerating the mails. It would have been a gross and manifest dereliction to have permitted that vitally important concern, the transportation of the mail—a concern so anxiously intrusted by the constitution to the federal authority—it would have been, in the opinion of the committee, a gross and manifest dereliction to have permitted it to lag behind the improvements of the age, and to be outstripped by the pace of ordinary travel and commercial communication. Such is the view which the Post Office Department avowedly takes of its own obligations, and upon which it habitually acts. To be outstripped by private expresses, or by the ordinary lines of travel, is deemed discreditable to the department, injurious to the general interests of the country, and a thing, therefore, not to be permitted.

This great and fundamental principle upon which the departments acts, (of not being outstripped in the transmission of correspondence and intelligence,) led necessarily to subsidizing the steam-engine into the service of the post office; and it must and will lead, with equal certainty, to a like adoption of any other newly discovered agency or contrivance possessing decided advantage of celerity over previously used methods. It is not probable, however, that the government will ever find itself called upon to make any transition wider or more striking than that already so familiar to us—a transition from the use of animal power to the tremendous enginery of the steam-engine; from common roads to iron railways; from land carriage to the conversion of rivers, lakes, and the ocean itself, into post roads.

The same principle which justified and demanded the transference of the mail on many chief routes, from the horse-drawn coach on common highways to steam-impelled vehicles on land and water, is equally potent to warrant the calling of the electro magnetic telegraph—that last and most wondrous birth of this wonder-teeming age—in aid of the post office, in discharge of its great function of rapidly transmitting correspondence and intelligence. And the only question to be considered, in determining whether it ought to be so called in aid, is a question of fact—namely, whether said telegraph possesses, over the modes of transmission now in use by the department, any advantages of sufficient value to justify the expense of engraftment on the system.

Its first and most signal advantage consists in the truly electrical celerity with which it transmits intelligence and communications through the greatest distances. It supplies, with a perfection like magic, the first and most important and difficult *desideratum* in a post office establishment—especially in one which has to serve a country so vast as ours. That desideratum is despatch—rapidity of transmission. It is to secure this, that the government pays a hugely greater price for the carrying and delivery of the mails, than any other equal *quantum* of transportation costs in the world. Nature seemed to have fixed certain limits to the speed of transmission, which it seemed impossible to pass; and those limits appeared to be reached by the steam-engine. But they have been utterly transcended by the electro magnetic telegraph, which has literally demolished time and space for all purposes of correspondence between places connected by its wonder-working wires.

Another inestimably important advantage of Professor Morse's telegraph consists in the fulness, precision, and variety of matter which it is capable of communicating. Its alphabet contains representatives of all the letters of our language, and of all the numerals of arithmetic; and they are capable of infinite combination and repetition under the magnetic impulse. Hence it is obvious that the *capacity* of the instrument is competent to the communication of a long discourse of the greatest variety of thought and expression. But, as the telegraph letters must necessarily be despatched along the wire, and marked down, one by one, at the station to which they are transmitted, it is obvious that a long discourse must occupy considerable time, although the letters follow each other in the most rapid succession.

This brings the attention of the committee to a very material point, namely: the quantum of matter, or amount of intelligence, which the instrument would be capable of transmitting in a given time. The ordinary average of transmission is about thirty letters per minute along each wire. Six wires can be erected at an expense of somewhat less than $500 per mile, which would make the telegraph competent to the transmission of one hundred and eighty letters per minute, on an average. The words of our language are estimated to average six letters to a word. A telegraphic line composed of six wires, would, consequently, be able to transmit per minute thirty words fully spelt. But it is wholly unnecessary that the words should be fully spelt by the instrument. By a well-contrived system of abbreviations, the number of letters to be transmitted, in order to communicate a given number of words, is greatly diminished; and, of course, the number of words transmissible in a given time is proportionably augmented. To such great perfection has this system of conventional abbreviations been carried, as to have enabled the telegraph, on one occasion, to transmit in thirty minutes, from Washington to Baltimore, congressional intelligence enough to fill a column of the Baltimore Patriot. This was done, too, with only one wire. Increase the number of wires to six, as proposed in the bill introduced by the Committee on Commerce, and it follows that the capacity of the instrument will be adequate to the transmission of six long newspaper columns of matter in half an hour. Then it is to be further noted, that the telegraph is capable of working throughout the whole twenty-four hours, without intermission—in darkness as well as in daylight—in stormy weather as well as in serene—which would enable it to communicate in a single day two hundred and eighty-eight long newspaper columns of matter. All these facts put together, evince that the capacity of the instrument, in reference not only to the celerity of its communications, but in reference also to the kind and quantity of matter it can communicate in a given time, is such as to recommend it as a most efficient medium both of private correspondence and public intelligence.

That it is capable of being, and will actually be, at no distant day, extensively employed as such a medium, it seems to the committee there can be but little room to doubt. Such a result seems, indeed, to be rendered altogether certain, when, in addition to the capacities of the instrument, we take into consideration its cheapness. For little more than $100,000, Baltimore can be connected with New York; and for a like sum, New York with Boston. There would then be an unbroken telegraphic line from Boston to Washington; passing through New York, Philadelphia, Baltimore and the other considerable towns on the route. What a vast number of short commercial letters would such a line be able to attract to itself, and to despatch every day, far in advance of the ordinary transportation by mail. Nor would any danger of a detrimental divulgence of their affairs exist to deter merchants and men of business from resorting to the telegraph; because, in the first place, the simple expedient of a concerted cipher between distant correspondents would protect their communications with a shield of secrecy impenetrable even to the officers and managers of the telegraph. And in the next place, the very nature of their functions will require that these persons shall be men of great trustworthiness, and that they shall moreover be placed under the most stringent official obligations of secrecy in regard to the contents of private communications. Under such circumstances, men of business need no more apprehend danger of improper publicity from employing the telegraph, than from the necessity of having clerks in their counting-houses to pen and copy their correspondence.

If all these advantages should have the effect of attracting to the telegraph the amount of custom which to the committee seems probable, it is obvious that a very moderate tariff of charges would produce income enough to make it a gainful property—at least upon such a line as that from Boston to Washington. It is upon this ground the committee base the belief that it is destined soon to be established along that whole line, if not by government, certainly by private capital and enterprise; and then a state of things will immediately develop itself, which the people will never endure nor tolerate the government in permitting to exist. That state of things would be that the post office, in its transportation of all correspondence and news, would lag not hours, but days, behind the transmission of the same things through another medium; and that, a medium belonging to private individuals, and controlled by private views and interests.

The importance of prompt action in the matter on the part of the government is further apparent from the fact that the invention is a private patented property. It is a property to the production of which Professor Morse has devoted years of the highest order of labour—the labour of genius and science combined. Under the patronage and at the expense of the government, he has been enabled to give to the world, in the line between Baltimore and Washington, a visible and perfectly triumphant demonstration of the success and utility of his invention. But the pecuniary reward, to which he is so justly entitled, remains yet in abeyance. It depends upon his being successful in making contracts with the government, or others, for the use of his invention. And, of course, if government shall not speedily embrace the project, and enable him to realize a compensation for his discovery, he will be necessitated to look elsewhere for his indemnification and reward. And, should the arrangements into which he may find it necessary to enter with private individuals or associations, stipulate exclusive rights in their favour, it is manifest how greatly government and people would lie at their mercy. Having in their hands the monopoly of such a medium of intelligence on the important lines, they could make such use of their advantages over the government and the community as would at length enable them to exact their own terms as the price of the surrender of their exclusive right; for the truth cannot be too often repeated, or too deeply impressed in relation to the subject, that the people will never submit long to the mischiefs and discredit of the public post office transmission of correspondence and intelligence being outstripped by any private monopoly or establishment whatever. The loss of revenue will co-operate with the complaints and sufferings of the people to compel the government, in the long run, to do what were better done at once—namely, to establish the telegraph in connection with, and as a branch of, the post office, on such great lines of communication as the correspondence and commerce of the country may indicate.

An accident has occurred, during the present winter, in the administration of the post office, to which the committee beg leave to call the attention of the House, as illustrative of the principles and policy by which the Department avowedly feels itself bound, as the public organ for the transmission of correspondence and intelligence. It is well known that, from Boston to Covington, in Georgia, the great southern and New Orleans mail is transported wholly by steam-power, either on water or on rail roads. It is carried this whole distance in five days. From Covington to Mobile, it is carried in stages, with the exception of a short interval of rail road in approaching Montgomery, Alabama. At Mobile, the mail is again committed to steam-carriage, by sea, to New Orleans. Now, of so much importance was a single day's anticipation of the mail deemed in all the great cities on the route, that a private express was established with that view, to be carried on horseback between Covington and Montgomery. All matter destined for the private express was addressed to the agents of the company at Covington and Montgomery, according as such matter should happen to come from the north or south. The express carrier at Covington, receiving his despatches from the northern mail on the arrival of the steam-cars, delivered them at Montgomery to the post office again in such time that they were sent from Montgomery to Mobile and New Orleans by mail one day in advance of the other letters, which reached Covington at the same time. The effect was, the anticipation of the northern mail by one day at Mobile and New Orleans; and the same operation, from Montgomery to Covington, resulted in one day's anticipation of the Mobile and New Orleans mails at New York. What did the Post Office Department do, under this state of facts? The answer to this question is to be found on the records of the Department. The Postmaster General, after watching these anticipations for a short time, issued an order for their prevention by the establishment of a post office express between Covington and Montgomery, to run alongside the private express. In the same manner, if the government shall not soon establish the telegraph on public account on the great routes, it will find the mails anticipated in the hundred-fold greater degree by the establishment of private telegraphs, which it will have to meet, either by purchasing them out on exorbitant terms, or by erecting a rival public telegraph line by their side.

The facts and reasonings which have now been submitted, satisfy the committee that it is important that the government should lose no time in occupying, with a telegraph of its own, the ground between Baltimore and New York. The committee look to the probability that the line will afterwards he progressively extended northwardly, southwardly, and westwardly, on routes the business and correspondence on which shall justify and require telegraphic facilities of communication.

Apart from the post office power, the government undoubtedly possesses the authority to establish the telegraph for its own use in the transmission of official orders and communications. On this ground, as well as on that growing out of the post office power, the committee deem the constitutionality of the measure incontrovertible.

The committee might easily add to the views and arguments which they have now presented, others of a highly commanding character—especially those which relate to the extreme value of which the magnetic telegraph would be in the emergencies of war, and its singular adaptedness to render our system of government easily and certainly maintainable over the immense space from the Atlantic to the Pacific, which our territory covers. Doubt has been entertained by many patriotic minds how far the rapid, full, and thorough intercommunication of thought and intelligence, so necessary to a people living under a common representative republic, could be expected to take place throughout such immense bounds. That doubt can no longer exist. It has been resolved and put an end to forever by the triumphant success of the electro magnetic telegraph of Professor Morse, as already tested by the government.

The fact that a bill has been long pending in the House, introduced by the Committee on Commerce, for the extension of the telegraph from Baltimore to New York, renders it unnecessary for this committee to report a bill. Without pronouncing positively on the sufficiency of the provisions of that bill, the committee consider the whole subject worthy the prompt attention of Congress.

Having thus presented their views on the subject referred to them, the committee beg to be discharged from its further consideration.

HISTORY OF TELEGRAPHS,

Employing Electricity in Various Ways for the Transmission of Intelligence.

We presume it will not be uninteresting to the reader, to be presented with an account of the various discoveries, in their chronological order, by which the science of Electricity became known to the world during the seventeenth and eighteenth centuries, and *prepared the way* for those more magnificent results, which have been made in this the nineteenth century. We will endeavour to make it as brief as is consistent with the importance of the subject, to enable us to mark the succession of discoveries and improvements through two hundred years.

More than any other branch of experimental philosophy, that of electricity had been most neglected, until the seventeenth century. The attractive power of amber is mentioned by Theophrastus and Pliny, and also later by others.

[16] In the year 1600, William Gilbert, a native of Colchester, and a London physician, published a Latin Treatise, *De Magnete*, in which he relates a variety of electrical experiments. He increased the list of electric bodies and also of substances upon which electrics could act, and noted some of the circumstances relating to their action. His theory of electricity was, however, very imperfect.

In 1630, Nicolaus Cabœus at Terrara, repeated Gilbert's experiments and made some progress, increasing the list of electrics; as also did Mr. Boyle in the year 1670. He made some discoveries which had escaped the observation of those who preceded him. Cotemporary with Mr. Boyle, Otto Guericke, burgomaster of Magdeburg, (the inventor of the air pump,) made some advances. He constructed a sulphur globe, which he mounted upon an axis, in a wooden frame, and causing it to revolve, at the same time rubbing the globe with his hand, performed a variety of electrical experiments. He was the first to discover, that a body once attracted by an excited electric, was repelled by it, and not again attracted until it had touched some other body. He observed the light and sound produced by the electric fluid, while turning his electrical machine. Dr. Wall about the same time observed the light and sound produced by rubbing pieces of amber with wool, and also experienced a slight shock. He compared the sound and light of the electric fluid so produced, to thunder and lightning.

Sir Isaac Newton also engaged in similar electrical experiments, and gave an account of them to the Royal Society in 1675. Mr. Hauksbee, whose writings are dated 1709, distinguished himself by experiments and discoveries in electrical attraction, and repulsion, and electric light. He constructed an electrical machine, adopting the glass, instead of the sulphur globe. He experimented upon the subtilty and copiousness of the electric light, and likewise upon the sound and shocks produced by the fluid. After the death of Mr. Hauksbee, the science of electricity made but slow progress, and few experiments were made for twenty years. In the year 1728, Mr. Stephen Grey, a pensioner at the Charter House, commenced his experiments with an excited glass tube. He and his friend, Mr. Wheeler, made a great variety of experiments in which they demonstrated, that electricity may be communicated from one body to another, even without being in contact, and in this way, may be conducted to a great distance. Mr. Grey afterwards found, that, by suspending rods of iron by silk or hair lines, and bringing an excited tube under them, sparks might be drawn, and a light perceived at the extremities in the dark. He electrified a boy suspended by hair lines; and communicated electricity to a soap bubble blown from a tobacco pipe. He electrified water, contained in a dish, placed upon a cake of rosin, and also a tube of water. He made some curious experiments upon a small cup of water, over which, at the distance of an inch, he held the excited tube. He observed the water to rise in a conical shape, from which proceeded a light; small particles of water were thrown off from the cone, and the tube moistened.

Mr. Du Fay, intendant of the French king's gardens, repeated the experiments of Mr. Grey in 1733. He found that by wetting the pack-thread he succeeded better with the experiment of communicating the electric virtue through a line 1256 feet in length. He made the discovery of two kinds of electricity, which he called *vitreous* and *resinous*; the former produced by rubbing glass, and the latter from excited sulphur, sealing wax, &c. But this he afterwards gave up as erroneous. Mr. Grey, in 1734, experimented upon iron rods and gave rise to the term *metallic conductors*. He gave the name *pencil* of *electric light* to the stream of electricity, such as is seen to issue from an electric point. He suggested the idea that the electric virtue of the excited tube was similar to that of thunder and lightning, and that it could be accumulated.

Dr. Desaguliers commenced his experiments in 1739. He introduced the term *conductor* to that body to which the excited tube conveys its electricity. He called bodies in which electricity may be excited by rubbing or heating, *electric per se*; and *non-electric* when they receive electricity, and lose it at once upon the approach of another non-electric. In the year 1742, several Germans engaged in this subject. Mr. Boze, a professor at Wittemburg, revives the use of Hauksbee's globe, instead of using Grey's glass tube, and added to it a *prime conductor*. Mr. Winckler substituted a cushion instead of the hand, which had before been employed to excite the globe. Mr. P. Gordon, a Benedictine monk and professor of philosophy at Erford, was the first who used a *cylinder* instead of a globe. With his electrical machine he conveyed the fluid through wires 200 ells in length and killed small birds. Dr. Ludolf of Berlin, in the year 1744, kindled by electricity the *ethereal spirit* of Frobenius, by the excited glass tube; the spark proceeding from an iron conductor. Mr. Boze fired gunpowder by electricity. Mr. Gordon contrived the electrical star. Mr. Winckler contrived a wheel to move by the agency of the same fluid. Mr. Boze conveyed electricity from one man to another by a jet of water, when both were placed upon cakes of rosin, six paces apart. Mr. Gordon fired spirits, by a jet of water; and the Germans invented the electrical bells.

Mr. Collinson in 1745 sent to the Library Company of Philadelphia, an account of these experiments, together with a tube, and directions how to use it. Franklin, with some of his friends, immediately engaged in a course of experiments, the results of which are well known. He was enabled to make a number of important discoveries, and to propose theories to account for various phenomena, which have been universally adopted, and which bid fair to endure for ages.

In the year 1745, such was the attention given to the subject of electricity, that experiments upon it were publicly advertised and exhibited for money in Germany and Holland. Dr. Miles, of England, in the same year fired phosphorus by the application of the excited tube itself without the intervention of a conductor. It was at this period that Dr. Watson's attention was given to this subject. He fired air, made inflammable by a chemical process, and discharged a musket by the electric fluid. He made many experiments, some of which will be described as we proceed.

The year 1745 was made famous by the discovery of the *Leyden Phial* by Mr. Cuneus a native of Leyden. It appears also to have been discovered by Mr. Von Kleist, dean of the Cathedral in Camin about the same time. By this discovery, electricity could be accumulated and severe shocks given. Mr. Gralath, in 1746, gave a shock to twenty persons at once, and at a considerable distance from the machine. He constructed the electrical battery by charging several phials at once. Mr. Winckler, and also M. Monnier, in France, transmitted the electric fluid through several feet of water as a part of the circuit. M. Nollet, in France, killed birds and fishes by the discharge of the Leyden jars. Improvements were made by Dr. Watson, and others, in the Leyden phial, by coating the inside and outside of it with tin foil. Abbé Nollet gave a shock to 180 of the guards in the king's presence; and at the grand convent of the Carthusians in Paris, the whole community formed a line of 3600 feet in length, by means of wires between them. The whole company upon the discharge of the phial, gave a sudden spring at the same instant. The French philosophers tried the same experiment through a circuit of persons, holding wires between them, two and a half miles in length. In another experiment the water of the basin in the Tuilleries was made a part of the circuit.

M. Monnier, the younger, to discover the velocity of electricity, discharged the Leyden phial through an iron wire 4000 feet in length, and another 1319 feet, but could not discover the time required for its passage. Dr. Franklin communicated his observations, in a series of letters, to his friend Collinson, the first of which is dated March 28, 1747. In these he shows the power of points in drawing and throwing off the electrical matter. He also made the grand discovery of a *plus* and *minus*, or of a *positive* and *negative* state of electricity. Shortly after Franklin, from his principles of plus and minus state, explained, in a satisfactory manner, the phenomena of the Leyden phial. Dr. Watson and others in July 18, 1747, conveyed the electric fluid across the Thames at Westminster bridge; the width of the river making a part of the circuit. On the 24th of July, he tried the experiment of forcing the electric fluid to make a circuit with the bend of the river, at the New River at Stoke, Newington. He supposed that the electric fluid would follow the river alone, through its circuitous windings, and return by the wire. He suspected from the result of this experiment, that the ground also conducted the fluid. On the 28th, he proved the fact by supporting a wire 150 feet in length upon baked sticks, using the ground as half of the circuit. On the 5th, of August, he tried another experiment of making the *dry* ground a part of the circuit for a mile in extent, and found it to conduct equally as well as water. The last experiment was tried at Shooter's Hill, on the 14th of August of the same year. But one shower of rain had fallen for the five preceding weeks. The wires, two miles in length, were supported upon baked sticks, and the dry ground was used for the return two miles of the circuit. They found the transmission of the electric fluid to be instantaneous. Dr. Watson made many other experiments which we must pass over.

Mr. Ellicott constructed an electrometer for measuring the quantity of electricity. Mr. Maimbury, at Edinburgh, electrified two myrtle trees, during the month of October, 1746, when they put forth small branches and blossoms sooner than other shrubs of the same kind, which had not been electrified. The same experiment was tried upon seeds, sowed in garden pots with the same success. Mr. Jallabert, Mr. Boze and the Abbé Menon principal of the College of Bueil, at Angers, tried the same experiments upon plants, by electrifying bottles in which they were growing. He proved that electrified plants always grew faster, and had finer stems, leaves and flowers than those which were not electrified.

In the year 1748, Dr. Franklin, and his friends, held an *electrical feast*[17] on the banks of the Schuylkill near Philadelphia, and as the account is amusing, as well as scientific, we will give an account of it as related by Franklin, in a letter to his friend Collinson, dated Philadelphia, 1748. (1 vol. of Franklin's Works, p. 202.)

"Chagrined a little that we have been hitherto able to produce nothing in this way of use to mankind; and the hot weather coming on, when electrical experiments are not so agreeable, it is proposed to put an end to them for this season, somewhat humorously, in a party of pleasure, on the banks of the *Skuykil*."

"Spirits, at the same time, are to be fired by a spark sent from side to side through the river, without any other conductor than the water, an experiment which we some time since performed, to the amazement of many. A turkey is to be killed for our dinner by the *electrical shock*, and roasted by the *electrical jack*, before a fire kindled by the *electrified bottle*: when the healths of all the famous electricians of England, Holland, France, and Germany are to be drank in *electrified bumpers*,[18] under a discharge of guns from the *electrical battery*."

"In the year 1749, Franklin first suggested his idea of explaining the phenomena of thunder gusts, and of the aurora borealis, upon electrical principles. He points out many particulars in which lightning and electricity agree; in the same year he conceived the bold idea of ascertaining the truth of his doctrine, by actually drawing down the lightning, by means of sharp pointed iron rods, raised into the region of the clouds. Admitting the identity of electricity and lightning, and knowing the power of points in repelling bodies charged with electricity, and in conducting the fluid silently and imperceptibly, he suggested the idea of securing houses, ships, &c. from being damaged by lightning, by raising pointed rods several feet above the most elevated part of the building to be protected, and the other end descending some feet into the ground. It was not until the summer of 1752, that he was enabled to complete his grand discovery by experiments."

"While he was waiting for the erection of a spire, it occurred to him that he might have more ready access to the region of clouds, by means of a common kite. He prepared one by fastening two cross sticks to a silk handkerchief, which would not suffer so much from the rain as paper. To the upright stick was affixed an iron point. The string was, as usual, of hemp, except the lower end, which was silk. Where the hempen string terminated, a key was fastened. With this apparatus, on the appearance of a thunder gust approaching, he went out into the commons, accompanied by his son, to whom alone he communicated his intentions, well knowing the ridicule which, too generally for the interests of science, awaits unsuccessful experiments in philosophy. He placed himself under a shade, to avoid the rain; his kite was raised—a thunder cloud passed over it—no sign of electricity appeared. He almost despaired of success, when, suddenly, he observed the loose fibres of his string to move towards an erect position. He now presented his knuckle to the key, and received a strong spark; repeated sparks were drawn from the key; a phial was charged, a shock given, and all the experiments made which are usually performed with electricity."

"Franklin constructed rods so as to bring the lightning into his house, for the purpose of ascertaining if it was of the positive or negative kind. He succeeded in the experiment for the first time in April, 1753, when it appeared that the electricity was negative. On the 6th of June he met with a cloud electrified positively. The discoveries of Franklin roused the attention of all Europe, and many distinguished electricians repeated them with success. Professor Richman, of St. Petersburg, while making some experiments upon the electrical state of the atmosphere, was killed by the electric fluid, August, 1753. Towards the end of the eighteenth century, electricity was assiduously cultivated by a great number of eminent individuals, who extended the boundaries of the science by numerous experiments, and by the invention of ingenious and useful instruments. Experiments were made upon air, water and ice; and in relation to the surfaces of electric bodies; in relation to the two electrical states; upon the deflagration of the metals; decomposition of solids and liquids," &c. &c.

Lomond's Electrical Telegraph.

It is stated in Young's Travels in France, (1787, 4th ed. vol. 1, p. 79,) that a Mr. Lomond had invented a mode by which, from his own room, he held communication with a person in a neighbouring chamber, by means of electricity. He employed the common electrical machine placed at one station, and at the other an electrometer constructed with pith balls. These instruments were connected by means of two wires stretched from one apartment to the other; so that, at each discharge of the Leyden phial, the pith balls would recede from each other, until they came in contact with the return wire. His system of telegraphic correspondence is not related. We must suppose from the character of his invention, having but one movement, that of the divergence of the balls, and using an apparatus extremely delicate, that his means of communication could not have been otherwise than limited, and required a great amount of time.

The only mode in which it appears possible for him to have transmitted intelligence, seems to be this: a single divergence of the pith balls, succeeded by an interval of two or three seconds, may have represented A. Two divergencies in quick succession, with an interval following, may have represented B; three divergencies, in like manner, indicated the letter C; and so on for the remainder of the alphabet. Instead of these movements of the pith balls representing letters, they may have indicated the numerals 1, 2, 3, &c. so that with a vocabulary of words, numbered, conducted his correspondence. This appears to be the first electrical telegraph of which we have any account; but does not appear to have been used upon extended lines.

Reizen's Electric Spark Telegraph.

In 1794, according to Voigt's Magazine, vol. 9, p. 1, Reizen made use of the electric spark for telegraphic purposes. His plan was based upon the phenomenon which is observed when the electric fluid of a common machine is interrupted in its circuit by breaks in the wire, exhibiting at the interrupted portions of the circuit a *bright spark*. The spark thus rendered visible in its passage he appears to have employed in this manner.

Fig. 34.

Figure 34 is a representation of the table upon which were arranged the letters of the alphabet, twenty-six in number. Each letter is represented by strips of tin foil, passing from left to right, and right to left, alternately, over a space of an inch square upon a glass table. Such parts of the tin foil are cut out, as will represent a particular letter. Thus, it will be seen that the letter A is represented by those portions of the tin foil which have been taken out, and the remaining portions answer as the conductor. P and N represent the positive and negative ends of the strips, as they pass through the table and reappear, one on each side of the small dot at A. Those two lines which have a dot between, are the ends of the negative and positive wire belonging to one of the letters. Now if a spark from a charged receiver is sent through the wires belonging to letter A, that letter will present a bright and luminous appearance of the form of the letter A. "As the passage of the electric fluid through a perfect conductor is unattended with light, and as the light or spark appears only where imperfect conductors are thrown in its way, hence the appearance of the light at those interrupted points of the tin foil; the glass upon which the conductors are pasted, being an imperfect conductor. The instant the discharge is made through the wire, the spark is seen simultaneously at each of the interruptions, or breaks, of the tin foil, constituting the letter, and the whole letter is rendered visible at once." This table is placed at one station, and the electrical machine at the other, with 72 wires inclosed in a glass tube connecting the two stations. He could have operated with equal efficiency by using 37 wires having one wire for a common communicating wire, or with 36 wires by substituting the ground for his common wire. It does not appear that it was ever tested to any extent.

Dr. Salva's Electric Spark Telegraph.

In 1798, Dr. Salva, in Madrid, constructed a similar telegraph, as that suggested by Reizen, (see Voigt's Magazine, vol. 11, p. 4.) The Prince of Peace witnessed his experiments with much satisfaction, and the Infant Don Antonio engaged with Dr. Salva in improving his instruments. It is stated that his experiments were conducted through many miles. No description of his plans appear to have been given to the public.

Origin of Galvanism.

Galvanism takes its name from Galvani, Professor of Anatomy at Bologna, who discovered it in the year 1790. As the account of the circumstances attending the discovery of this useful and wonderful agent, may not be uninteresting to the reader, we insert it here as related in the "*Library of Useful Knowledge.*"

"It happened in the year 1790, that his wife, being consumptive, was advised to take, as a nutritive article of diet, some soup made of the flesh of frogs. Several of these animals, recently skinned for that purpose, were lying on a table in the laboratory, close to an electrical machine, with which a pupil of the Professor was amusing himself in trying experiments. While the machine was in action, he chanced to touch the bare nerve of the leg of one of the frogs with the blade of the knife that he held in his hand; when suddenly the whole limb was thrown into violent convulsions. Galvani was not present when this occurred, but received the account from his lady who had witnessed, and had been struck with the singularity of the appearance. He lost no time in repeating the experiment: in examining minutely all the circumstances connected with it, and in determining those on which its success depended. He ascertained that the convulsions took place only at the moment when the spark was drawn from the prime conductor, and the knife was at the same time in contact with the nerve of the frog. He next found that other metallic bodies might be substituted for the knife, and very justly inferred that they owed this property of exciting muscular contractions to their being good conductors of electricity. Far from being satisfied with having arrived at this conclusion, it only served to stimulate him to the farther investigation of this curious subject; and his perseverance was at length rewarded by the discovery, that similar convulsions might be produced in a frog, independently of the electrical machine, by forming a chain of conducting substances between the outside of the muscles of the leg, and the crural nerve. Galvani had previously entertained the idea, that the contractions of the muscles of animals were in some way dependent on electricity; and as these new experiments appeared strongly to favour this hypothesis, he with great ingenuity applied it to explain them. He compared the muscles of a living animal to a Leyden phial, charged by the accumulation of electricity on its surface, while he conceived that the nerve belonging to it, performed the function of the wire communicating with the interior of the phial, which would, of course, be charged negatively. In this state, whenever a communication was made by means of a substance of high conducting power between the surface of the muscle and the nerve, the equilibrium would be instantly restored, and a sudden contraction of the fibres would be the consequence.

"Galvani was thus the first to discover the reason of that peculiar convulsive effect which we now obtain from the Galvanic battery, and he attributed it to a modification of electricity. It was left to another to construct an instrument which would give a constant and increased effect, and develop this extraordinary fluid. Whatever share accident may have had in the original discovery of Galvani, it is certain that the invention of the Pile, an instrument which has most materially contributed to the extension of our knowledge in this branch of physical science, was purely the result of reasoning.

"Professor Volta, of Pavia, in 1800, was led to the discovery of its properties by deep meditation on the developements of electricity at the surface of contact of different metals. We may justly regard this discovery as forming an epoch in the history of galvanism; and since that period, the terms Voltaism, or Voltaic electricity, have been often, in honour of this illustrious philosopher, used to designate that particular form of electrical agency.

"He had been led by theory to conceive that the effect of a single pair of metallic plates might be increased, indefinitely, by multiplying their number, and disposing them in pairs, with a less perfect conducting substance between each pair. For this purpose he provided an equal number of silver coins, and of pieces of zinc, of the same form and dimensions, and also circular discs of card, soaked in salt water, and of somewhat less diameter than the metallic plates. Of these he formed a pile or column as shown in figure 35, in which three substances, silver, zinc, and wet card, denoted by the letters S, Z, I, were made to succeed one another in the same regular order throughout the series. The efficacy of this combination realized the most sanguine anticipations of the discoverer. If the uppermost disc of metal in the column be touched with the finger of one hand, previously wetted, while a finger of the other hand is applied to the lowermost disc, a distinct shock is felt in the arms, similar to that from a Leyden phial, or still more nearly resembling that from an electrical battery, weakly charged. These discs are supported by two large discs, *a* and *i*, of wood, one at the bottom and the other at the top of the pile, with three glass rods, A, B, C, at equal distances around the pile, but not touching it, and are cemented into the wooden base and cover. P represents the wire connecting the silver disc, and N that connecting the zinc."

Fig. 35.

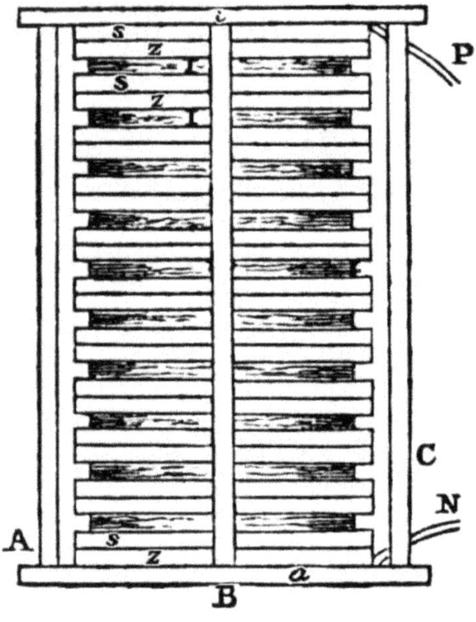

The Decomposition of Water.

"The chemical agency of galvanism, exerted on *fluid* conductors, placed in the circuit between the poles of the battery, is very remarkable. Among the simplest of its effects is the resolution of water into its two gaseous elements, oxygen and hydrogen. The discovery of this fact is due to the united researches of Mr. Nicholson and Mr. Carlisle, and was one of the immediate consequences of the invention of the pile by Volta. The most convenient mode of exhibiting the decomposition of water by the Voltaic battery, is to fill, with water, a glass tube; to each end of which, a cork has been fitted so as to confine the water, and to introduce into the tube two metallic wires, by passing one, at each end, through the cork which closes it, allowing the extremities of the wires, that are in the water, to come so near each other as to be separated by an interval of only a quarter of an inch. The wires being then respectively made to communicate with each of the two poles of a Voltaic battery, the following phenomena will ensue. If the wire connected with the positive pole of the battery consists of an oxidable metal, it is rapidly oxidated by the water surrounding it; while, at the same time, a stream of minute bubbles of hydrogen gas arises from the surface of the other wire, which is in connection with the negative pole. But if we employ wires made of a metal which is not susceptible of oxidation by water, such as gold or platina, gas will be extricated from both the wires, and, by means of a proper apparatus may be collected separately."

We shall now see that these two discoveries, viz. the Voltaic pile, and the decomposition of water by the agency of the former are the bases of a plan *for telegraphic* purposes.

Samuel Thomas Soemmering's Description of his Voltaic Electric Telegraph, invented in 1809.

Fig. 36.

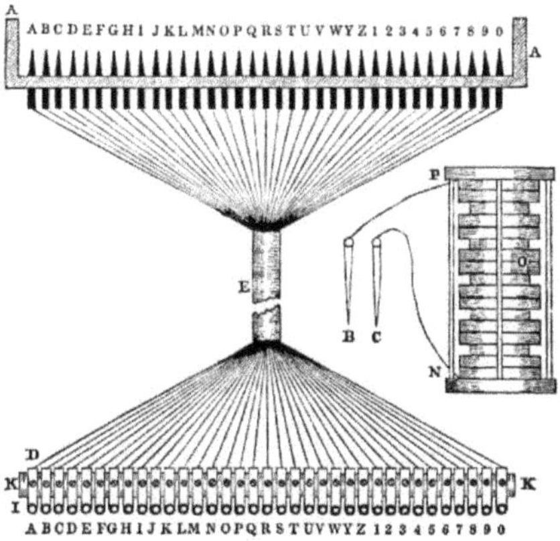

"The fact that the decomposition of water may be produced with certainty and instantaneously, not only at short, but at great distances from the Voltaic pile, and that the decomposition may be sustained for a considerable time, suggested to me the idea, that it might be made subservient for the purposes of transmitting intelligence in a manner superior to the plan in common use, and would supersede them. My engagements were such that I have only been able to test the practicability of my plan upon a small scale, and herewith submit, for the Academy's publication, an account of the experiment.

"My telegraph was constructed and used in the following manner: In the bottom of a glass reservoir, figure 36, of which A A is a sectional view, are 35 golden points, or pins, passing up through the bottom of the glass reservoir, marked A, B, C, &c. 25 of which are marked with the 25 letters of the German alphabet and the ten numerals. The 35 points are each connected with an extended copper wire, soldered to them, and extending through the tube, E, to the distant station; are there soldered to the 35 brass plates, upon the wooden bar, K K. Through the front end of each of the plates, there is a small hole, I, for the reception of two brass pins, B and C; one of which is on the end of the wire connecting the positive pole, and the other the negative pole of the Voltaic column, O. Each of the 35 plates are arranged upon a support of wood, K K, to correspond with the arrangement of the 35 points at the reservoir, and are lettered accordingly. When thus arranged, the two pins from the column are held, one in each hand, and the two plates being selected, the pins are then put into their holes and the communication is established. Gas is evolved at the two distant corresponding points in an instant. For example, K and T. The peg on the hydrogen pole, evolves hydrogen gas, and that on the oxygen pole, oxygen gas.

"In this way every letter and numeral may be indicated at the pleasure of the operator. Should the following rules be observed, it will enable the operator to communicate as much if not more, than can be done by the *common telegraph.*

"*First Rule.* As the hydrogen gas evolved is greater in quantity than the oxygen, therefore, those letters which the former gas represents, are more easily distinguished than those of the latter, and must be so noted. For example, in the words *ak*, *ad*, *em*, *ie*, we indicate the letters *A*, *a*, *e*, *i*, by the hydrogen; *k*, *d*, *m*, *e*, on the other hand, by the oxygen poles.

"*Second Rule.* To telegraph two letters of the same name, we must use a unit, unless they are separated by the syllable. For example, the name *anna*, may be telegraphed without the unit, as the syllable *an*, is first indicated and then *na*. The name *nanni*, on the contrary, cannot be telegraphed without the use of the unit, because *na* is first telegraphed, and then comes *nn*, which cannot be indicated in the same vessel. It would, however, be possible to telegraph even three or more letters at the same time by increasing the number of wires from 25 to 50, which would very much augment the cost of construction and the care of attendance.

"*Third Rule.* To indicate the conclusion of a word, the unit 1 must be used. Therefore, it is used with the last single letter of a word, being made to follow the ending letter. It must also be prefixed to the letter commencing a word, when that letter follows a word of *two letters* only. For example: *Sie lebt* must be represented *Si*, *e1*, *le*, *bt*, that is the unit 1 must be placed after the first *e*. *Er lebt*, on the contrary, must be represented. *Er*, *1l*, *eb*, *t1*; that is, the unit 1 is placed before the *l*. Instead of using the unit, another signal may be introduced, the cross † to indicate the separation of syllables.

"Suppose now the decomposing table is situated in one city, and the pin arrangement in another, connected with each other by 35 continuous wires, extended from city to city. Then the operator, with his Voltaic column and pin arrangement at one station, may communicate intelligence to the observer of the gas at the decomposing table of the other station.

"The metallic plates with which the extended wires are connected have conical shaped holes in their ends; and the pins attached to the two wires of the Voltaic column are likewise of a conical shape, so that when they are put in the holes, there may be a close fit, prevent oxidation and produce a certain connection. It is well known that slight oxidation of the parts in contact will interrupt the communication. The pin arrangement might be so contrived as to use permanent keys, which for the 35 plates or rods would require 70 pins. The first key might be for hydrogen A; the third key for hydrogen B; the fourth key for oxygen B, and so on.

"The preparation and management of the Voltaic column is so well known, that little need be said except that it should be of that durability as to last more than a month. It should not be of very broad surfaces, as I have proved, that six of my usual plates (each one consisting of a Brabant dollar, felt, and a disc of zinc, weighing 52 grains) would evolve more gas, than five plates of the great battery of our Academy.[19] As to the cost of construction, this model which I have had the honour to exhibit to the Royal Academy, cost 30 florins. One line consisting of 35 wires, laid in glass or earthen pipes, each wire insulated with silk, making each wire 22,827 Parisian feet, or a German mile, or a single wire of 788,885 feet in length, might be made for less than 2000 florins, as appears from the cost of my short one."

Extract from the Journal of the Franklin Institute, vol. 20, page 325.

"To the foregoing notice, we append an article published in Thompson's Annals of Philosophy, vol. 7, page 162, 1st series, February, 1816. This article is from the pen of Dr. John Redman Coxe, of Philadelphia, and it is believed that the idea of the employment of galvanism, for a telegraph which it suggests, was then original. Those who are acquainted with the history of the progress of electricity, as evolved by the ordinary machine, are aware that experiments had been made with a view to its employment for a similar purpose; but from the inherent difficulties of the subject, the project had been abandoned.

"It is not pretended, that the state of our knowledge on the subject of galvanism, was such at the time the foregoing suggestion was made, as would have enabled any person to apply it practically; this, if done, will be due to the recent discoveries on the subject of electro magnetism; a subject which has been very successfully pursued by the philosophers of our own country, and particularly by Professor Henry, of Princeton. As some of the philosophers of Europe are disputing upon the question of the authorship of *proposition* for the employment of Galvanic electricity, telegraphically, we have thought that it would not be altogether inopportune, or uninteresting, to publish the article above referred to.

"*Use of Galvanism as a Telegraph: in an extract of a Letter from Dr. J. Redman Coxe, Professor of Chemistry, Philadelphia.*

"I observe in one of the volumes of your Annals of Philosophy, a proposition to employ galvanism, as a solvent, for the urinary calculus, but which has been very properly, I think, opposed by Mr. Armiger. I merely notice this, as it gives me the opportunity of saying, that a similar idea was maintained in a thesis, *three years* ago, by a graduate of the University of Pennsylvania. I have, however, contemplated this important agent, as a probable means of establishing telegraphic communications, with as much rapidity, and perhaps less expense, than any hitherto employed. I do not know how far experiment has determined galvanic action, to be communicated by means of wires; but there is no reason to suppose it confined, as to limits, certainly not as to time. Now, by means of apparatus, fixed at certain distances, as telegraphic stations, by tubes, for the *decomposition* of *water*, and of metallic salts, &c. regularly ranged, such a key might be adopted as would be requisite to communicate words, sentences, or figures, from one station to another, and so on to the end of the line, I will take another opportunity to enlarge upon this, as I think it might serve many useful purposes; but like all others, it requires time to mature. As it takes up little room, and may be fixed in private, it might, in many cases, of besieged towns, &c. convey useful intelligence, with scarcely a chance of detection by the enemy. However fanciful in speculation, I have no doubt that sooner or later, it will be rendered useful in practice."

"I have thus, my dear sir, ventured to encroach upon your time, with some crude ideas, that may serve to elicit some useful experiments in the hands of others. When we consider what wonderful results have arisen from the first trifling experiments of the junction of a small piece of silver and zinc in so short a period, what may not be expected from the further extension of galvanic electricity: I have no doubt of its being the chiefest agent, in the hands of nature, of the mighty changes that occur around us. If the metals are compound bodies, which I doubt not, will not this active principle combine those constituent in numerous places, so as to explain their metallic formation? and if such constituents are in themselves aeriform, may not galvanism reasonably tend to explain the existence of metals in situations to which their specific gravities certainly do not entitle us to look for them?"

Ronald's Electric Telegraph, invented in 1816.

From the Encyclopedia Britannica, 7th edition, page 662.

"M. Cavællo suggested the idea of conveying intelligence by passing a given number of sparks through an insulated wire in given spaces of time; and some German and American authors have proposed to construct galvanic telegraphs by the decomposition of water. Mr. Ronalds, who has devoted much time to the consideration of this form of the telegraph, proposes to employ common electricity to convey intelligence along insulated and buried wires, and he proved the practicability of such a scheme, by insulating eight miles of wire on his lawn at Hammersmith. In this case the wire was insulated in the air by silk strings. But he also made the trial with 525 feet of buried wire; with this view he dug a trench four feet deep, in which he laid a trough of wood two inches square, well lined within and without with pitch; and within this trough were placed thick glass tubes, through which the wire ran. The junction of the glass tubes was surrounded with shorter and wider tubes of glass, the ends of which were sealed up with soft wax.

"Mr. Ronalds now fixed a circular brass plate, figure 37, upon the second arbour of a clock which beat dead seconds. This plate was divided into twenty equal parts, each division being worked by a figure, a letter, and a preparatory sign. The figures were divided into two series of the units, and the letters were arranged alphabetically, omitting J, Q, U, W, X and Z. In front of this was fixed another brass plate as shown in figure 38, which could be occasionally turned round by the hand, and which had an aperture like that shown in the figure at V, which would just exhibit one of the figures, letters and preparatory signs, for example, 9, *v*, and ready. In front of this plate was suspended a pith ball electrometer, B, C, figure 38, from a wire D, which was insulated, and which communicated on one side with a glass cylinder machine, and on the other side with the buried wire. At the further end of the buried wire, was an apparatus exactly the same as the one now described, and the clocks were adjusted to as perfect synchronism as possible.

Fig. 37.

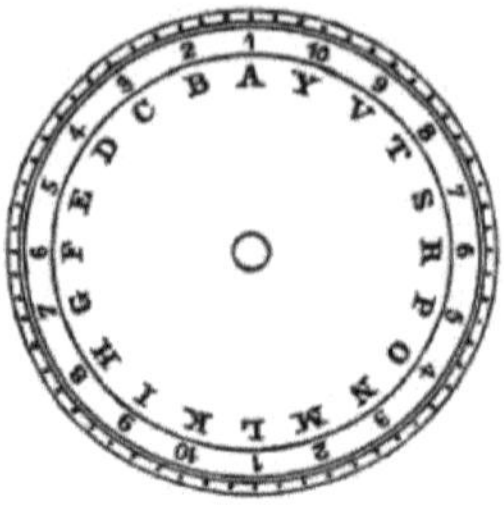

Fig. 38.

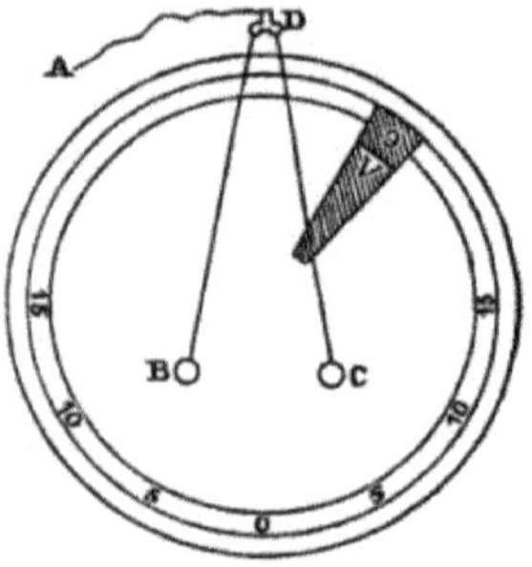

"Hence it is manifest, that when the wire was *charged* by the machine at either end, the electrometers at both ends *diverged*, and when it was discharged, they collapsed, at the same instant. Consequently, if it was discharged at the moment when a given letter, figure, and sign on the lower plate, figure 37, appeared through the aperture, figure 38, the same figure, letter and sign would appear also at the other clock; so that by means of such discharges at one station, and by marking down the letters, figures and signs, seen at the other, any required words could be spelt.

"An electrical pistol was connected with the apparatus, by which a spark might pass through it when the sign *prepare* was made, in order that the explosion might excite the attention of the superintendent, and obviate the necessity of close watching.

"*Preparatory signs.* A, prepare; V, ready; S, repeat sentence; P, repeat word; N, finish; L, annul sentence; I, annul word; G, note figures; E, note letters; C, dictionary."

Electro Magnetism.

We have now to notice a discovery, which forms the basis of those modern telegraphs in which the principle of electro magnetism is adopted. The following is an extract from the "Library of Useful Knowledge," in relation to the discovery:

"The real discoverer of the magnetic properties of electric currents M. Oersted, Professor of Natural Philosophy, and Secretary of the Royal Society of Copenhagen. In a work which he published in the German, about the year 1813, on the identity of chemical and electrical forces, he had thrown out conjectures concerning the relations subsisting between the electric, galvanic and magnetic fluids, which he conceived might differ from one another only in their respective degrees of tension. If galvanism, he argued, be merely a more latent form of electricity, so magnetism may possibly be nothing more than electricity in a still more latent form; and he, therefore, proposed it as a subject worthy of inquiry, whether electricity employed in this, its most latent form, might not be found to have a sensible effect upon a magnet. It is difficult clearly to understand what he meant by the expression of *latent states*, as applied to electricity, but it may be sufficient for us to know, that in the various endeavours he subsequently made to verify his conjectures, he was led to such forms of experiment as afforded decisive indications of the influence of Voltaic currents on the magnetized needle. Yet, even after he had succeeded thus far, it was a matter of extreme difficulty to determine the real direction of this action, and it was not till the close of the year 1819, that his perseverance was at length rewarded by complete success.

"The first account of his discovery that appeared in England is contained in a paper, which he himself communicated, in Thompson's Annals of Philosophy, for October, 1820, vol. 16, page 273; and in which the following experiments are described. The two poles of a powerful Voltaic battery were connected by a metallic wire, so as to complete the galvanic circuit. The wire which performs this office he called the *uniting* wire; and the effect, whatever it may be, which takes place in this conductor, and in the space surrounding it, during the passage of the electricity, he designates by the term *electric* conflict, from an idea that there takes place some continued collision and neutralization of the two species of electric fluids, while circulating in opposite currents in the apparatus. Then taking a magnetic needle, properly balanced on its pivot, as in the mariner's compass, and allowing it to assume its natural position in the magnetic meridian, he placed a straight portion of the uniting wire horizontally above the needle, and in a direction parallel to it; and then completed the circuit, so that the electric current passed through the wire. The moment this was done, the needle changed its position, its ends deviating from the north and south towards the east and west, according to the direction in which the electric current flowed, so that by reversing the direction of the current the motion of the needle was also reversed. The general law he expressed as follows: 'That end of the needle which is situated next to the negative side of the battery, or towards which the current of positive electricity is following, immediately moves to the westward.'

"The deviation of the needle is the same, whether the uniting wire, instead of being immediately above the needle, be placed somewhat to the east or west of it, provided it continue parallel to and also above it. This shows that the effect is not the result of a simple attractive or repulsive influence, for the same pole of the magnetic needle which approaches the uniting wire, when placed on its east side, recedes from it when placed on its west side."

[20] "Soon after this important discovery of Oersted's was made, M. Ampère established the second fundamental law of electro magnetism, that the two conducting wires from the poles of the battery, when conveniently

suspended, *attracts each other when they transmit electrical currents moving in the same direction, and repel each other when the currents which they transmit have opposite directions.*

"On the 25th Sept. 1820, M. Arago communicated to the French Institute the important discovery that the electrical current possesses, in a very high degree the power of developing magnetism in iron or steel. Sir H. Davy communicated a similar fact to Dr. Wollaston on the 12th of November, 1820, and Dr. Seebeck laid before the Royal Academy of Berlin a series of experiments on the same subject.

"M. Arago found that the uniting wires of a powerful Voltaic battery attracts iron filings often with such force as to form a coating around the wire ten or twelve times thicker than itself. This attraction, as he found, did not originate in any magnetism previously possessed by the iron filing, which he ascertained would not adhere to iron, and that it was not a case of common electrical attraction, was evident from the fact that copper and brass filings were not attracted by the uniting wire. M. Arago likewise found, that the iron filings began to rise before they came in contact with the uniting wire; and hence he drew the conclusion, that the electric currents converted each small piece of iron into a temporary magnet. In following out this view, the French philosopher converted large pieces of iron into temporary magnets and also small steel needles into permanent ones, (by employing the helix.) Sir H. Davy and Dr. Seebeck obtained analogous results without knowing what had been previously done in France.

"A galvanometer was first constructed by Professor Schweigger, of Halle, very soon after the first discovery of electro magnetism, and by him called an *electro* magnetic multiplier."

In the year 1820, Ampère predicted the possibility of making the deflection of the magnetic needle, by the agency of the galvanic fluid, serve the purposes of transmitting intelligence. In page 19 of his memoir, he thus resolves the problem:

"As many magnetic needles as there are letters of the alphabet," he says, "which may be put in action by conductors; which may be made to communicate successively with the battery by means of keys; which may be pressed down at pleasure, might give place to a telegraphic correspondence which would surmount all distance and would be as prompt as writing speech to transmit thought."

"The next step in the progress of discovery, was that of making magnets of extraordinary power by means of a galvanic battery. This seems to have been first accomplished by Prof. Moll, of Utrecht, and Professor Henry, of

Princeton, who was able to lift thousands of pounds weight by his apparatus."

The following Extract is taken from a Work on Electro Magnetism published by Jacob Green, M. D. Professor of Chemistry in Jefferson Medical College, 1827.

"In the very early stage of electro magnetic experiments, it had been suggested, that an instantaneous telegraph might be constructed by means of conjunctive wires, and magnetic needles. The details of this contrivance are so obvious, and the principles on which it is founded so well understood, that there was only one question which could render the result doubtful. This was, whether by lengthening the conjunctive wires, there would be any diminution in the electrical effect upon the needle. It is the general opinion, that the electrical fluid, from a common electrical battery, may be transmitted, without any sensible diminution, instantaneously, through a wire three or four miles in length. At the philosophical dinner, as it has been called, got up a number of years ago by some gentlemen of Philadelphia, on the banks of the Schuylkill, it may be recollected that Dr. Franklin killed a turkey with the electric shock, transmitted across the river, a distance of more than half a mile; and Dr. Watson, who was also at the pains of making some experiments of this kind, asserts that the electric shock was transmitted, instantaneously, through the length of 12,276 feet. Had it been found true that the galvanic fluid could be transmitted in a moment through a great extent of conducting wire, without diminishing its magnetic effect then no question could have been entertained as to the practicability and importance of the suggestion adverted to above, with regard to the telegraph. Mr. Barlow, of the Royal Military Academy, who has made a number of successful experiments and investigations in electro magnetism, fully ascertained that there was so sensible a diminution with only 200 feet of wire, as to convince him at once of the impracticability of the scheme.

Triboaillet's Proposition.

[21] "In 1828, M. Victor Triboaillet de Saint Amand proposed to establish a correspondence from Paris to Brussels, by placing along the highway, and at some feet deep, a metallic wire, about a line or a line and a half diameter. He recommended to cover the wire with shellac, upon which was to be wound silk, very dry, which should be covered in their turn with a coating of resin. The whole was then to be put into glass tubes carefully luted up with a resinous substance and secured by a last envelope in the earth, then varnished over and hermetically sealed. Then, by means of a powerful battery, he would communicate the electricity to the conducting wire, which would transmit

the current to the opposite point to an electroscope, destined to render sensible the slightest influence, and left to each one to adopt at pleasure the number of motions to express the words or letters which they might need."

Fechner's Suggestion.[22]

"Fechner, in his manual of galvanism, (Voss, 1829, page 269,) remarked, that the electro magnetic effects of the galvanic current would be far more appropriate for the giving of signs than Soemmering's plan by the decomposition of water."

He suggested that wires, having twenty-four multiplicators should be extended between Leipsic and Dresden, and there connected, alternately, with a galvanic column, for telegraphic purposes. Indeed, he ventured to prophecy, that probably hereafter such a connection between the central point of a kingdom, and different provinces might be arranged as there was existing in animal bodies, between the central point of organic structure of particular members and nerves.

Magneto Electricity.

We come now to give an account of a new branch in the science of electricity, viz. *magneto electricity*; which Dr. Faraday was the first to discover in the year 1831. As this species of electricity has been applied to several of the plans of electric telegraphs, which we shall describe, it is desirable that some account should be given of its discovery, and of the instrument by which it is generated.

The following is an extract from "Daniell's Introduction to Chemical Philosophy" 2d edition, London, 1843.

"The phenomena of electro magnetism are produced by *electricity in motion*; accumulated electricity, when *not in motion*, exerts no magnetic effects. Dr. Faraday early felt convinced that "as every electric current is accompanied by a corresponding intensity of magnetic action at right angles to the current, good conductors of electricity, when placed within the sphere of this action, should have a current induced through them, or some sensible effect produced, equivalent in force to such a current." These considerations, with their consequence, the hope of obtaining electricity from ordinary magnetism, stimulated him to investigate the subject experimentally, and he was rewarded by an affirmative answer to the question proposed. He thus became, like Oersted, the founder of an entirely new branch of natural philosophy.

"If a wire connecting the two ends of a delicate galvanometer be placed parallel and close to the wire connecting the poles of a Voltaic battery, no effect will be produced upon the needle, however powerful the current may be. If the points opposed in the two wires be multiplied by coiling the one, as a helix, *within the convolutions* of the other, coiled in the same way, both being covered with silk to prevent metallic contact, still no effect will be discernible so long as the current is *uninterrupted.* When, however, the current of the battery is stopped by breaking the circuit, the needle is momentarily deflected, as by a wave of electricity passing in the same direction as that of the main current. Upon allowing the needle to come to a state of rest, and then renewing the contact, a similar impulse will be given to it in the contrary direction. While the current continues, the needle returns to its state of rest, again to be deflected in the first direction by stopping the current. Motion may be accumulated to a considerable amount in the needle, by making and breaking the contacts with the battery in correspondence with its swing. The same effects are produced when, the current being uninterrupted, the conducting wire is made suddenly to approach or recede from the wire of the galvanometer. As the wires approximate, there will be a momentary current induced in the direction contrary to the inducing current; and as the wires recede, an induced current in the same direction as the inducing current.

"As this *Volta electric induction* is obviously produced by the transverse action of the Voltaic current, in one case, by the *mechanical* motion of the wire, and in the other at the moments of *generation* and *annihilation* of the current, Dr. Faraday thought that the sudden induction and cessation of the same magnetic force in soft iron, either by the agency of a Voltaic current, or by that of a common magnet, ought to produce the same results. He constructed a combination of helices (8) upon a hollow cylinder of pasteboard: they consisted of lengths of copper wire, containing, altogether, 220 feet; four of these were connected end to end, and then with the galvanometer. The other intervening four were also connected end to end, and then with the Voltaic battery. In this form a slight effect was produced upon the needle by making and breaking contact. But when a soft iron cylinder, seven-eighths of an inch thick and twelve inches long, was introduced into the pasteboard tube, surrounded by the helices, the induced current affected the galvanometer powerfully. When the iron cylinder was replaced by an equal cylinder of copper, no effect beyond that of the helices alone was produced.

"Similar effects were then produced by *ordinary magnets*. The hollow helix had all its elementary helices connected with the galvanometer, and the soft iron cylinder having been introduced into its axis, a couple of bar magnets were arranged with their opposite poles in contact, so as to resemble a horse-shoe magnet, and contact was then made between the other poles and the ends of the iron cylinder, by which it was converted, for the time, into a magnet; by breaking the magnetic contacts, or reversing them, the magnetism of the iron cylinder could be destroyed or reversed at pleasure. Upon making magnetic contact, the needle was deflected; continuing the contact, the needle became indifferent, and resumed its first position; on breaking contact, it was again deflected, but in the opposite direction to the first effect, and then it again became indifferent. When the magnetic contacts were reversed, the deflections were reversed. The actual contacts of the magnets with the soft iron is not essential to the success of these experiments, for their near approximation induces sufficient magnetism in the cylinder to generate the electric current, which affects the needle. The first rise of the magnetic force induces the electric wave in one direction; its sudden decline, in the opposite. Mechanical motion of a permanent magnetic pole in one direction, across the coils of the helix, will produce the same effect as the sudden induction of the magnetism in the soft iron, and its motion in the opposite direction will cause a corresponding effect with its annihilation, when the soft iron cylinder is removed from the helix, and one end of a cylindrical magnet thrust into it, the needle is deflected in the same way as if the magnet had been formed, by either of the two preceding processes. Being left in, the needle will resume its first position, and then being withdrawn, the needle will be deflected in the opposite direction. On substituting a small hollow helix, formed round a glass tube, for the galvanometer, in these experiments, and introducing a steel needle, it will be converted into a magnet, provided care be taken not to expose it to the opposite action of the reverse current; and if the continuity of the conducting wire be broken, at the moment when the secondary electric wave is passing through it, a bright spark may be obtained.

"The connection of electro magnetical and magneto electrical phenomena may be exhibited in a very striking way, by employing any of the apparatus, by which the rotary motions of the *magnet* or *conducting wire*, are produced by a current of electricity, to generate electric currents by the mechanical rotations of the magnet or wire. For this purpose, the galvanometer may be substituted for the battery, and when the wire is made to turn round the pole of the magnet, or the pole of the magnet round the wire, in one direction, the needle will be deflected to one side; and to the other by the opposite rotation. Nothing can be better shown that *magneto electric* is the *converse* of *electro* magnetic action.

"Dr. Faraday by rotating a copper disc between the poles of a horse-shoe magnet, produced a constant current of electricity in one direction, and deflected the needle of the galvanometer; one wire being connected with the disc, and the other with the arbour. By turning the disc in one direction, the circuit will pass from the axis to the circumference; by turning it in the opposite direction, the current will flow from the circumference to the axis."

Fig. 39.

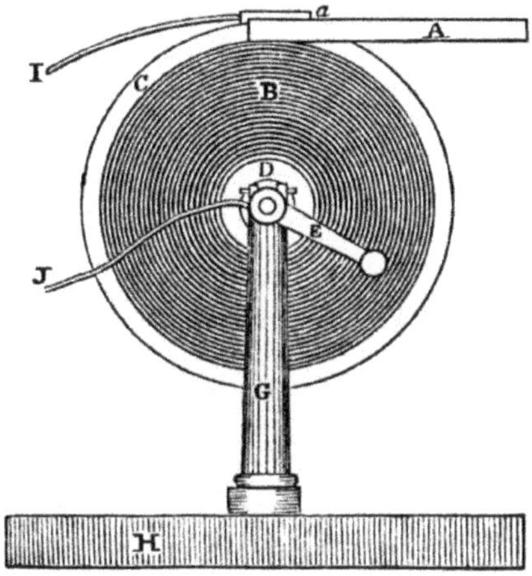

Figure 39 represents a side view of the instrument. B shows the copper disc permanently secured upon its axis, and which is turned by means of the crank, E. G represents one of the standards which support the axis. H is the platform upon which the various parts are arranged. The edge, C, of the copper disc, is amalgamated so as to make a perfect connection with the amalgamated segment, *a*, to which is soldered a wire, I, leading to the galvanometer. That portion of the disc, B, which is shaded, is not amalgamated. J is the other wire proceeding from the galvanometer, and both it and the axis are amalgamated, at the points of connection. A is the permanent magnet, with its poles on each side of the copper disc, B, opposite the amalgamated portion of the rim.

Fig. 40.

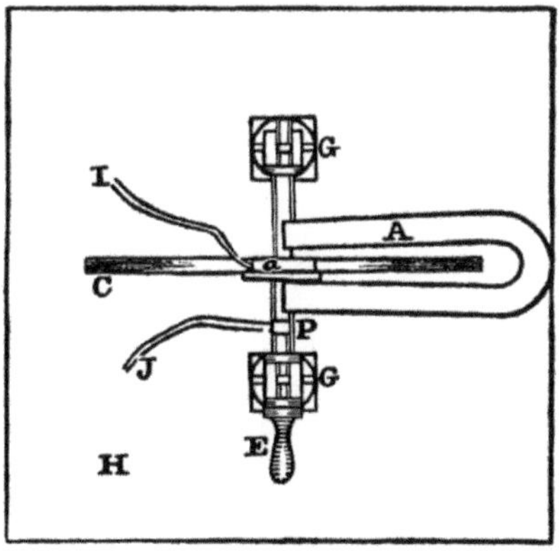

Figure 40, represents a top view of the instrument, H is the platform; C the disc; *a* the segment; A the permanent magnet; J the wire attached to the axis,

P; G and G are the two standards. E the crank; and I the wire attached to the segment *a*.

Mr. Saxton,[23] in a letter to Mr. Lukens, dated, London, April 14th, 1832, after describing Dr. Faraday's rotating disc, figures 39 and 40, says, "I have made this experiment in a different way, and succeeded satisfactorily. The method was as follows: A coil of wire wrapped with silk, similar to that used in the galvanometer, was attached, by the ends, to the wires of the galvanometer. On passing this roll, backward and forward, upon one of the poles of a horse-shoe (permanent) magnet, or placing it upon and removing it from either pole, I have made the needle of the galvanometer to spin round rapidly." Figure 41, represents Mr. Saxton's plan.

Fig. 41.

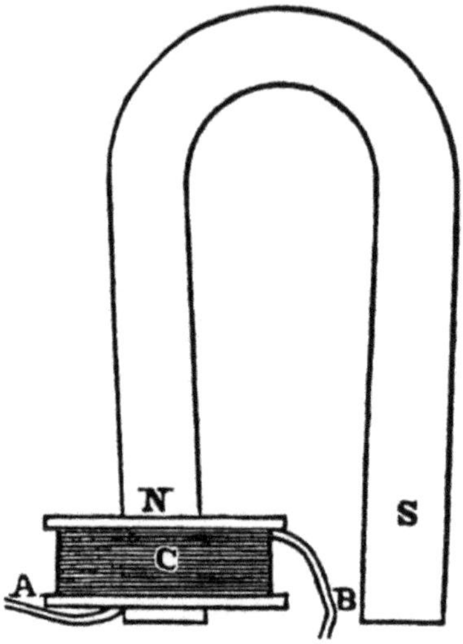

N and S represent the north and south poles of the horse-shoe permanent magnet. C is the coil of wire, wound round a spool of an oblong shape, through the centre of which there is an opening sufficiently large to admit either of the prongs of the magnet through it. A and B are the ends of the wire leaving the coil, and are connected with the galvanometer.

Mr. Saxton on the 2d of May, 1832, obtained the spark by the following arrangement of the permanent magnet and the helix of wire round the armature. In relation to this instrument, he thus writes to Mr. Lukens, of Philadelphia, dated, London, May 11th, 1832. Jour. Frank. Int. vol. 13, p. 67. "Since my last I have heard of a method of producing a spark from a magnet, discovered I think by an Italian.[24] This experiment I made at once upon a large horse-shoe magnet, which I am making for Mr. Perkins and his partners. One of your large magnets will answer the same purpose. Make a cylinder of soft iron of an inch, or three-fourths of an inch, in diameter, and of the usual length of the keeper; place two discs of brass or wood upon this cylinder, and at such a distance apart that they will conveniently pass between the poles of the magnet; between these wind, say fifty feet of bobbin wire, which may be of iron covered with cotton; let the ends of this coil be bent over the ends of the cylinder and brought down until they touch the poles of the magnet. The ends should be of such a length, that on bringing the cylinder to the magnet, one of the ends will touch, when the cylinder is about half an inch from the magnet, and the other at one-fourth of an inch. The cylinder being thus arranged, and in contact with the magnet, on drawing it suddenly away a spark will pass between the end of the wire, and the pole of the magnet."

Fig. 42.

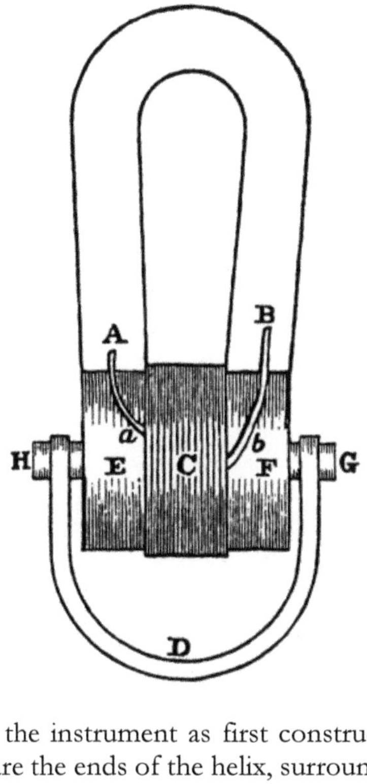

Figure 42 represents the instrument as first constructed by Mr. Saxton, in London.[25] A and B are the ends of the helix, surrounding the cylindrical bar of soft iron between E and F, filling the cavity which has been formed out of the solid iron. The size of bar between the collars E and F, thus formed, is the same as the projections H and G. The wire, *a*, proceeds from the outside of the coil and makes a suitable contact upon the prong, A, of the magnet: *b* proceeds from the bottom of the coil, where the winding commenced and makes a similar contact upon the prong, B, of the permanent magnet. One wire extends a little further upon the magnet than the other, so that the shorter one may break its connection sooner than the longer. H and G are projections from the sides of the armature, to which the handle, D, is secured. Let the armature, with its helix, be held up against the ends of the prongs of the permanent magnet; and the wires *a* and *b*, in perfect contact with their respective prongs, as shown in the figure; if, while in this condition, the keeper is suddenly withdrawn, a spark will appear at the end of the short wire, as it breaks its contact with the prong of the magnet.

Mr. Saxton, however, was still further successful, the following year, in carrying out an idea which occurred to him on the 6th of December, 1832, of producing the same phenomena, with a more convenient and powerful rotating instrument.[26] This new arrangement he was able to test on the 20th of June, 1833, and obtained the spark. On the 22d, he made an unsuccessful attempt, in the presence of Prof. Rogers, of Philadelphia, at the decomposition of water. On the 30th of June he exhibited it at a meeting of the British Association at Cambridge, before Dr. Faraday, Dr. Brewster, Prof. Forbes, Dr. Dalton, and many other distinguished and scientific gentlemen. The experiments made by it were the exhibition of the spark, giving shocks, &c. On the third of July, Mr. Saxton succeeded in decomposing water, by adding a little sulphuric acid, and on the 25th of August, he ignited and melted platinum wire.

Fig. 43.

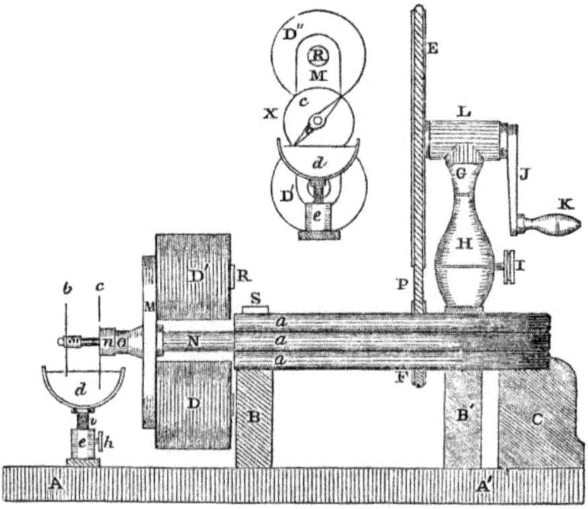

Figure 43 exhibits a side view of the instrument: *a*, *a*, *a*, is a compound permanent magnet, consisting of three steel plates, put together, side by side. B and C are two wooden supports, upon the platform A A. To these supports the magnet is permanently secured, by a yoke, S, through which pass two screws into the wooden supports below. M is a cross bar, into which, and at right angles with it, are screwed two arms of round soft iron, R, about five-eighths of an inch in diameter, the whole forming the armature or keeper of the magnet. Upon these two projecting arms, are placed two coils, D' and D, of copper wire, insulated with silk. The whole is very securely fastened upon the steel spindle, N, which has its journals in the supports, B and B'. On the end of the spindle, N, near the curve of the magnet, there is a small band pulley, F, which is driven by the band or cord of the large wheel, E, and the crank, J. The axis of the large wheel passes through a long socket, L, in the top of the column, H; on one end of the axis the band wheel is fastened and on the other the crank. By this arrangement a very rapid and quiet rotary motion is given to the armature.

In the column, H, there is a socket, into which the stem of the upper part of the column, G, is fitted, which admits of the large wheel being raised or lowered, so as to prevent the band from slipping, and when properly adjusted it is secured by the screw, I. O is an ivory hub, sliding over that part of the spindle immediately projecting beyond the cross bar, M. Upon this ivory hub is a copper disc, C, with a socket, *n*: *b*, is a needle made of platinum, which, with its socket, *m*, is nicely fitted upon the end of the steel spindle, so as to be adjusted to any required angle with the armature, and when adjusted to retain its position. The two ends of the two coils, which leave the *centre* of the helices, are made to form a contact with the soft iron arms, R R, passing through the coils, D' and D, thus making the circuit complete with the needle *b*, upon the end of the spindle, N, by a continuous metallic connection of the arms, with the cross bar, M, and through the cross bar to the spindle, N, in contact with the needle, *b*. The two ends of the two coils, leaving the *outside* of the helices, are joined in one, and as they pass through the cross bar, M, are insulated from it, by a piece of ivory, inserted in the cross bar. The united wire then passes into and through the ivory hub, *e*, forming a perfect contact with the copper disc, *c*, underneath its socket, *n*: *d*, is a cup of mercury, in which the copper disc, *c*, is always immersed, and the needle, *b*, twice in every revolution of the armature. The cup, *d*, is so constructed as to rise and fall, by means of a stem, *i*, sliding vertically into a socket, *e*, of its support, and is secured to its position by the screw, *h*. In this way, its proper height for breaking and closing the circuit may be easily obtained, when the armature is rotating. The proper position for the needle, *b*, is that in which it is just leaving the mercury, as the keeper arrives at the position, in which its magnetism is neutralized. This position is seen at X, where D'' and D' are the sides of the coils; *c* the copper disc; *m* the cross bar of the armature; R the arm passing through the coil; and *b* the needle, at that *angle* which it requires, when the armature is vertical or at its neutral position. It will be observed that the needle is just leaving the mercury, *d*.

Fig. 44.

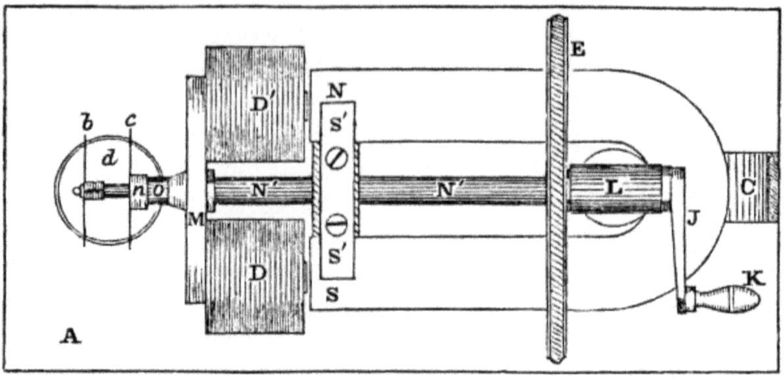

Figure 44 represents a top view. N and S represent the north and south poles of the permanent magnet. N' and N' is the spindle, parallel with the prongs of the magnet, and equidistant from them; L is the socket of the band wheel; D' and D the horizontal position of the coils; M is the cross bar; *b* the needle; *c* the copper disc, and *m* and *n* their respective sockets; *o* the ivory hub; *d* the cup of mercury; A the platform; and S' and S' the yoke through which pass two screws to secure the magnet to the wooden support below.

When the armature is made to rotate, it becomes a temporary magnet, by the laws of magnetic induction, whenever the arms carrying the helices come opposite to the poles of the permanent magnet, and when these soft iron arms have reached the point at *right angles* to the *magnet*, or vertical, their magnetism for an instant is destroyed, and are as instantaneously reversed from what they were before reaching that point. They are also magnetic, just in that proportion as they recede from or approach to the poles of the permanent magnet.

Hence, first, *one arm* is the south pole, when opposed to the north pole of the magnet; and the *other arm* a north pole, when opposed to the south pole of the magnet. But when they have made a half revolution on their axis, from their first position, their magnetism is reversed. The arm which was a south pole, has become a north pole; and the arm that was a north pole has become a south pole. Thus, by the rotation of the armature, direction of the induced current in the arms, become changed, as often as they are alternately brought opposite the poles of the permanent magnet, which is *twice* in every revolution of the armature.

It follows, then, by the laws of magneto induction, that as often as the arms become magnetic, they induce corresponding *opposite electric* currents in the wire surrounding those arms, provided the circuit of the coils is complete. The disc, which is in metallic connection with two ends of the wire leaving the coils, (one from each coil,) is *always* immersed in the mercury of the cup. The needle, however, which is in connection with the other two wires from the two coils, (one from each coil,) is *not always* immersed, but only when the armature is at a certain position in relation to the permanent magnet. The circuit then can only be closed when the needle is immersed, as well as the disc. Upon inspecting the figure, it will be found that the needle is immersed at the time the arms are passing the poles of the magnet, and that when they arrive at the vertical or neutral position, the needle has just broken its connection with the mercury, and at that instant the spark is observed.

Professor Daniell observes, that "by means of this magneto electrical machine, all the well known effects of Voltaic currents may be very commodiously produced. When the communication is made between the spindle and the revolving disc, by means of a fine platinum wire, instead of the dipping points, the wire may be maintained at a red heat; although the effect being produced by alternating currents in opposite directions, a kind of pulsation, or intermission of the light, may be discerned. Upon making the communication between the two mercury cups, by means of copper cylinders *grasped* in the *hands*, a continued painful contraction of the muscles of the arm takes place, which destroys voluntary motion, and, under certain circumstances, is perfectly *intolerable*.

"The general expression of these phenomena may be thus stated: whenever a piece of metal is passed, either before a single pole, or between the opposite poles of a magnet, or before electro magnetic poles, whether ferruginous or not, so as to cut the magnetic curves, (or lines, which would be marked out by a spontaneous arrangement of iron filings,) electrical currents are produced across the metal, transverse to the direction of motion."

Dr. Page's Magneto Electric Machine.

This important instrument also depends, for its action, upon the principle discovered by Dr. Faraday, that electricity was developed in conducting bodies, when they were moved in a certain direction, in the neighbourhood of permanent magnets. Since the beautiful and ingenious invention which Mr. Saxton was the first to make, no valuable improvements have been made in this machine, except those introduced by Professor Page.

The first important change in the machine, was the adaptation of his pole changer to the machine, in place of the break pieces, which were used in all the modifications up to that time; and another equally useful improvement, consisted in the arrangement of the permanent magnets and armatures. Previous to this last improvement, these machines were constructed with a single permanent magnet, and one or more revolving armatures, necessarily involving great disadvantages. Page's improvements were completed in

February, 1838, and shortly after published in Silliman's Journal. He was also the first to suggest the combination of several machines under one mechanical movement, as the best mode of augmenting power in this way.

The combined machine, described in Daniell's Introduction to Chemical Philosophy, as invented by Wheatstone, about two years since, is the same as that described, and represented by Dr. Page in Silliman's Journal in 1838. In the same publication, Dr. Page described the arrangement of the permanent magnets and armatures, as shown in the annexed figures. The adaptation of the pole changer, which, in connection with this machine, is called the *Unitrep*, Dr. Page has given to the public. But as he has never allowed the improvement, which consists in the use of two or more permanent magnets and straight armatures, to be sold with his knowledge and consent, he intends to claim a *patent for the same*; it having been decided by our courts, that the publication of an invention by the inventor, does not affect his right to a patent, provided he does not allow the invention to be sold and used.

The figures 45, 46 and 47, exhibit one of Page's machines with his early improvements.

Figure 45, is a side elevation of the machine.

Figure 46, is a top view.

Figure 47, are views of the revolving armatures and coils.

In Figure 45, representing a side view of the machine, B and B are the compound permanent steel magnets, composed of six bars each of the U form, mounted upon the brass pillars, P, P, P, P, which are fastened into the common platform of the whole machine. Through the platform there pass stout rods, R and R, and upwards through two brass straps, above the magnets, B and B. These straps or yokes secure the magnets from any motion by means of the screw nuts. A is a circular case of pasteboard, containing the armatures and coils. H is a band wheel surrounding the case, for mechanical connection with any source of power that may be used to keep the machine in motion. I' and I are two metallic studs, with an aperture passing vertically from the top, to the depth of an inch, for the reception of connecting wires, and then, by means of a screw at its side, to make a perfect contact. There are two other studs directly behind them. G and J are the two pulley wheels, with their band and crank, by which a rapid rotary motion is given to the armatures and coils. These pulleys are supported by the standard. From the bottom of the studs I' and I, as also from those directly behind them, proceed wires which are carried along below the platform, and pass up through it between the pilliar, P, and the revolving armatures, to the shaft; there being one on each side of the axis.

Fig. 45.

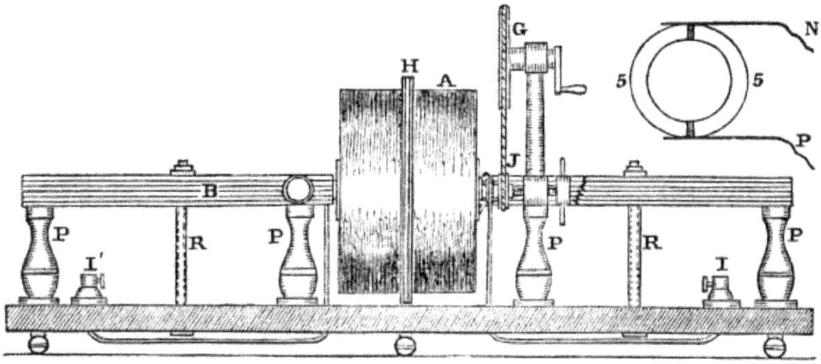

Fig. 46.

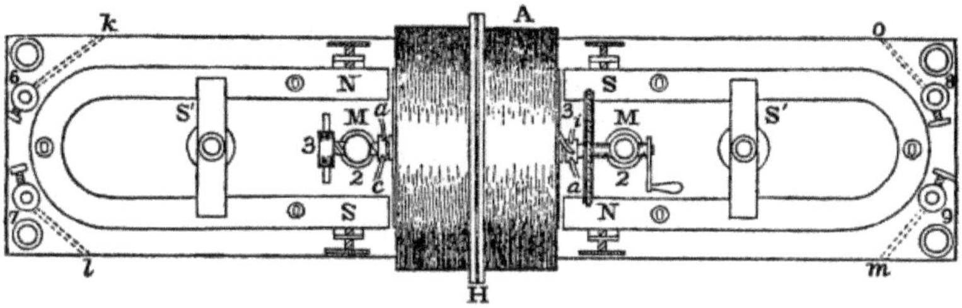

Figure 46, represents a top view of the instrument. A is the case containing the armatures and coils, and H the band wheel. N, S and N, S, are the north and south poles of the permanent magnets. S' and S' are the yokes by which the magnets are secured to the platform, and the screws near the poles of the magnets are for the purpose of setting the magnets to any required position, laterally, and securing them in it. M and M are the tops of the two pilliars, which support the shaft of the armatures and coils. The bearings are so made as to allow the apparatus to revolve with as little friction as possible: 3 and 3 represent the set screws against the ends of the shaft, for adjusting the ends of the permanent magnets; by which means, the armatures may be allowed to pass very near the ends of the magnets without touching. 6, 7, 8, and 9 are the receiving studs, by which the wires from any other instrument may be connected with the machine. The wire, *a*, in contact with the unitrep, as before stated, is continued and soldered to the receiving stud, 6; in the same manner, *c*, also in contact with the unitrep, is connected with 7; and also 3 with 8; and *a* with 9. The manner in which these wires, *a*, *c*, 3, and *a*, form their contact with the shaft, is seen at N and P, figure 45, of which 5 and 5 represent a section of the shaft and unitrep.

Fig. 47.

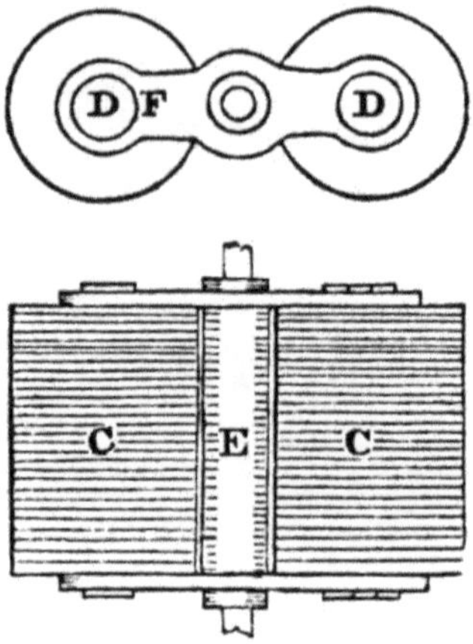

Figure 47 represents the revolving armatures and coils, with the case taken off. C and C are the two coils of insulated copper wire, surrounding two straight bars of soft iron, represented in the end view by D and D. E is the shaft. The two armatures and coils are secured to the two brass straps F, which are themselves fastened upon the shaft. The armatures are allowed to project through the straps about the sixteenth of an inch.

On each end of the shaft is attached an *unitrep*, consisting of two cylindrical segments of silver, as seen at 5 and 5, figure 45; insulated from each other, and secured to a cylinder of ivory or wood, upon the shaft, so as to revolve with it. The terminations of the coils of wire upon the armatures, are soldered to the segments of silver, and as the unitrep turns, it brings opposite ends of the wires, alternately, upon the stationary wires or conductors, P and N: (in figure 46 they are represented by *a* and *c*, and 3 and *a*.) The opposing currents of the coils, in each half revolution, are, by this contrivance, made to form one continuous current. Hence, the name *unitrep* (to turn together.) There being two unitreps, and corresponding conducting wires, and screw cups, the induced currents from the two coils may be combined in several ways, after the manner of combining separate batteries.

Let the wires below the base board be all properly connected with the receiving cups, as heretofore described. Then let the wire from 6, (represented by dots,) to *k*, be connected with the wire 9 and *m*; and also the wire 7 and *l*, with the wire 8 and 0. Let one of the united wires be connected with one wire leaving the coil of an electro magnet; and the other united wire be joined to the other wire of the electro magnet of the telegraph, or any other instrument designed to operate by a galvanic battery. When this preparation is finished, if the armatures and coils are made to revolve rapidly, a powerful current is formed in the induced coils, C and C, figure 47, capable of performing all the experiments generally made by means of the galvanic battery.

Dr. Page has made a very important discovery in connection with this machine, not now to be made known; but, suffice it to say, the single machine which he has now in his possession, on Christmas day, 1844, operated Morse's telegraph, through the circuit of 80 miles; half this circuit being wire, the other half the earth. This machine makes an electro magnet sustain 1000 pounds, and melts a platinum wire one-fortieth of an inch in diameter.

The Pole Changer.

We introduce here a description of an instrument used for reversing the direction of the galvanic current, and which is applied in the operation of several kinds of electric telegraphs. There is a variety of modes by which the same object is attained, but as this appears the most simple, we have chosen it in preference to others.

The following figures, 48, 49 and 50, are three views of the instrument as it appears when looking down upon it, in its three changes. First, that in which the current is broken and the needle vertical. Second, in which the circuit is closed and the needle deflected to the right. Third, in which the circuit is closed and the needle deflected to the left. Each figure has, in connection with the pole changer, the battery, or any other generator of the electric fluid, represented by N and P, and the galvanometer represented by G. In each of the figures, the circles numbered 1, 2, 3, 4, 5, 6, 7, and 8, represent cups, filled with mercury, let into the wood of the platform, and made permanent. The small parallel lines terminating in these cups, represent copper wires or conductors.

Fig. 48.

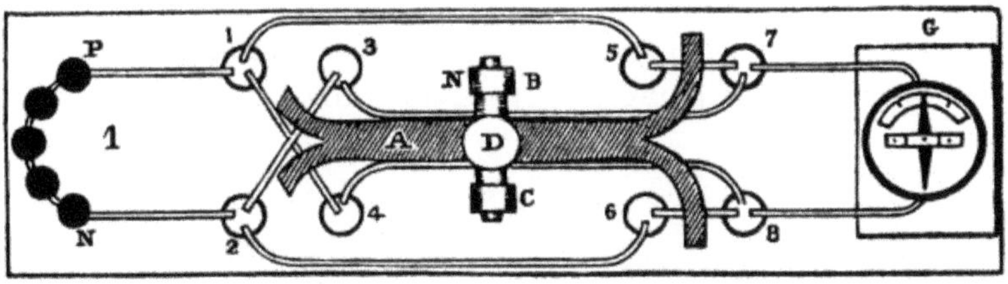

A, figure 48, represents a horizontal lever of wood, or some insulating substance, with its axis supported by two standards, B and C, by which it can easily vibrate. D represents an ivory ball, mounted upon a rod, inserted in the lever, and extending a few inches above it. It serves as a handle, by which to direct the elevation or depression of either end of the lever. Both ends of the lever branch out, presenting two arms each. Through each arm passes a copper wire, insulated from each other. The left hand branches support the wires which connect the mercury cups, 1 and 4, and 2 and 3, together. The right hand branches support the wires which connect the cups 5 and 7, and 6 and 8, together. The ends of these wires directly over the mercury cups are bent down, so that they may freely enter their respective vessels when required. The other wires are permanently secured to the platform. The position of the lever is now *horizontal*, and the bent ends of the wires, which it carries, are so adjusted, that none of them touch the mercury, consequently, there is no connection formed between the battery and galvanometer, and the needle is *vertical*. The ivory ball, it will be observed, is directly over the centre of the axis, and in that position required to break the circuit. Thus, the wires, 2 and 3, 1 and 4, 5 and 7, 6 and 8, are each out of the mercury, and the circuit being broken the fluid cannot pass.

Fig. 49.

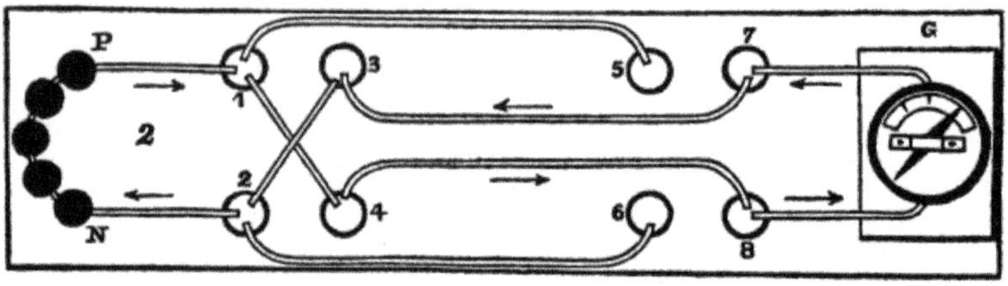

Figure 49 represents those connections which are formed when the left hand side of the lever is depressed, immersing in the mercury those wires supported by it. The ball and lever are omitted for the better inspection of the wires. Now the circuit is closed, and the current is passing from P, of the battery, to the mercury cup, 1; then along the cross wire to 4; to 8; to the coils of the multiplier, deflecting the needle to the *right*; then to 7; to 3; then along the cross wire, (which is not in contact with wire 1 and 4,) to 2; to the N pole of the battery. The arrows also show the direction of the current. It will be observed that the cups 5 and 7, and 6 and 8 are not now in connection, and consequently the current cannot pass along the wires 1 and 5, and 2 and 6.

Fig. 50.

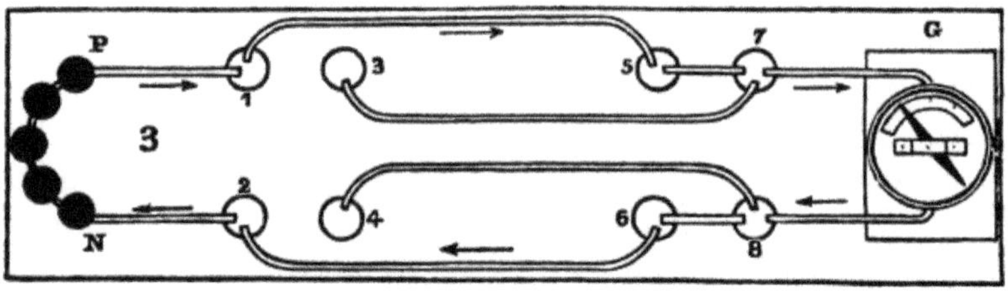

Now, if the ball, D, is carried to the right, a new set of wires, figure 50, are immersed, and those represented in figure 49, as in connection, are taken out of their cups. The fluid now passes from P, of the battery, to the mercury cup, 1; to 5; to 7; to the coils of the multiplier, deflecting the needle to the *left*; then it passes to cup 8; to 6; to 2; and then to the N pole of the battery; the arrows representing the direction of the current. It will now be found, that the cups, 2 and 3, and 1 and 4 are not in connection, and consequently the current cannot pass along the wires, 3 and 7, and 4 and 8.

Thus, it will appear, that by carrying the ball, D, to the left, the needle is deflected to the *right*; then, by carrying the ball to the right, the needle is deflected to the *left*; and that when the ball is brought to the vertical position, the needle is *vertical*. These three changes enter into the plans of several electric telegraphs, which are to be hereafter described.

Professor Morse's American Electro Magnetic Telegraph, invented, 1832.

To our readers the principles and arrangement of Morse's telegraph have been fully explained in the former part of this work. We shall here present *some* of the evidence of the time of its invention.

Extract from a letter from S. F. B. Morse to the Hon. Levi Woodbury, Secretary of the Treasury, dated Sept. 27th, 1837.

"About five years ago, on my voyage home from Europe, the electrical experiment of Franklin, upon a wire some four miles in length, was casually recalled to my mind, in a conversation with one of the passengers, in which experiment it was ascertained that the electricity travelled through the whole circuit in a time not appreciable, but apparently instantaneous. *It immediately occurred to me, that if the presence of electricity could be made* VISIBLE *in any desired part of this circuit, it would not be difficult to construct a* SYSTEM OF SIGNS *by which intelligence could be instantaneously transmitted.* The thought, thus conceived, took strong hold of my mind, in the leisure which the voyage afforded, and I planned a system of signs and an apparatus to carry it into effect. I cast a species of type, which I had devised for this purpose, the first week after my arrival home; and although the rest of the machinery was planned, yet, from the pressure of unavailable duties, I was compelled to postpone my experiments, and was not able to test the whole plan until within a few weeks. The result has realized my most sanguine expectations."

The following letters were published in the Journal of Commerce, from the originals now in possession of Prof. Morse.

Letter of the Hon. W. C. Rives.

SENATE CHAMBER, *September 21st, 1837.*

MY DEAR SIR,—I hope you will find in my multiplied and oppressive engagements here, an apology for not having sooner answered your inquiry on the subject of your Electro Magnetic Telegraph. I retain a distinct recollection of your having explained to me the conception of this ingenious invention, during our voyage from France to the United States in the year 1832, and that it was, more than once, the subject of conversation between us, in which I suggested difficulties which you met and solved with great promptitude and confidence.

I beg leave to assure you, that it would give us all great pleasure to renew, in personal intercourse at home, the agreeable souvenirs of our acquaintance, and friendly relations abroad.

I remain with great respect,
Your most obd't serv't,

W. C. RIVES.

PROF. S. F. B. MORSE.

Letter of Capt. William W. Pell,
of the packet ship Sully.

NEW YORK, *Sept. 27th, 1837.*

DEAR SIR—On my arrival here I received your letter, calling upon my recollection for what was said on the subject of an electric telegraph, during the passage from Havre, on board of the ship Sully, in October, 1832. I am happy to say, I have a distinct remembrance of your suggesting, as a thought newly occurred to you, the possibility of a telegraphic communication being effected by electric wires. As the passage progressed, and your idea developed itself, it became frequently a subject of conversation. Difficulty after difficulty was suggested as obstacles to its operation, which your ingenuity still labored to remove, until your invention, passing from its first crude state through different grades of improvement, was, in seeming, matured to an available instrument, wanting only patronage to perfect it, and call it into reality; and I sincerely trust that circumstances may not deprive you of the reward due to the invention, which, whatever be its source in Europe, is with you at least, I am convinced, original.

When you observed to me a few days before leaving the ship, "*well, Captain, when you hear of the telegraph one of these days, as the wonder of the world, remember the discovery was made on board the good ship Sully,*" I, then, little thought, I should ever be called upon to throw into the scale, my mite of testimony in support of your claims to priority of invention, for what seemed so startling a novelty.

With my respects and best wishes,
I subscribe myself,
WILLIAM W. PELL.

SAMUEL F. B. MORSE, ESQ.

A subsequent letter from Captain Pell, dated February 1st, 1838, after having seen the operation of the telegraph at the University, has the following paragraph:

"When, a few days since, I examined your instrument, *I recognized in it the principles and mechanical arrangements*, which, on board, I had heard you so *frequently explain* through all their developments."

From a letter now in possession of the author, and addressed to him by Prof. Morse, we make the following extract:

"In 1826, the lectures, before the New York Atheneum, of Dr. J. F. Dana, who was my particular friend, gave to me the first knowledge ever possessed of electro magnetism; and some of the properties of the *electro magnet*; a knowledge which I made available in 1832 as the basis of my own plan of an electro telegraph. I claim to be the original suggestor and inventor of the electric magnetic telegraph, on the 19th of October, 1832, on board the packet ship Sully, on my voyage from France to the United States, and, *consequently*, the inventor of the first, really *practicable telegraph on the electric principle.* The plan then conceived and drawn out in all its essential characteristics, is the one now in successful operation. All the telegraphs in Europe, which are practicable, are based on a different principle, and, *without an exception*, were invented subsequently to mine.

"The thought occurred to me, in a general conversation, as seated at the table with the passengers, in which the experiments of Franklin to ascertain the velocity of electricity through three or four miles. The thought at once occurred to me that electricity might be made the means of conveying intelligence, and that a system of signs might easily be devised for the purpose. I ought, perhaps, to say, that the conception of the idea of an *electric telegraph*, was original with me at that time, and I supposed that I was the first that had ever associated the two words together, nor was it until my invention was completed, and had been successfully operated through ten miles, that I, for the first time, learned, that the idea of an electric telegraph had been conceived by another. To me it was original, and its total dissimilarity from all the inventions and even suggestions of others, may be thus accounted for. I had not the remotest hint from others, till my whole invention was in successful operation. I employed myself in the wakeful hours of the night, as well as in the tedious hours of the day, *in devising the signs, adapting them to a single circuit of wire*, and in *constructing machinery which should record the signs upon paper*, for I thought of no plan short of a mode of recording."

On the second of September, 1837, the author, with several others, witnessed the first exhibition of this electric telegraph, and soon after became a partner with the inventor. Immediate steps were taken for constructing an instrument for the purpose of exhibiting its powers before the members of Congress. This was done at the Speedwell Iron Works, Morristown, N.J. and exhibited in operation with a circuit of two miles. A few days after, it was again exhibited at the University of the City of New York, for several days, to a large number of invited ladies and gentlemen. The circuit at this time was increased to ten miles. Immediately after this exhibition the instruments and ten miles of wire were taken to Washington, and continued in operation for several months, in the room of the Committee on Commerce at the capitol. Its history and progress, after this period, may be gathered from the preceding documents, printed by order of Congress.

Schilling Electric Telegraph.

We make the following extract in relation to Schilling's telegraph from the Polytechnic Central Journal, Nos. 31, 32, 1838:

"Baron Schilling, of Caunstadt, a Russian Counsellor of State, likewise occupied himself with telegraphs by electricity, (see Allgem Bauztg, 1837, No. 52, p. 440,) and had the merit of having presented a much simpler contrivance, and of removing some of the difficulties of the earlier plans. He reckoned many variations to the right, or left, following in a certain order for a telegraphic sign, as, indeed, in this manner, the needle was strongly varied, and only came to rest gradually, after many repeated vibrations; he introduced a small rod of platinum, with a scoop, which dipped into a vessel of quicksilver, placed beneath the needle, and by the check given, changed the vibration of the needle into sudden jerks. In order to apprise the attendant of a telegraphic despatch, he loosed an alarm. How much of this contrivance was Schilling's own, or whether a portion of it was not an imitation of Gauss and Weber, the author cannot decide, but that Schilling had already experimented, probably with a more imperfect apparatus, before the Emperor Alexander, and still later before Emperor Nicholas, is affirmed by the documents quoted."

From the report of the "Academy of Industry," Paris, February, 1839, we make the following extract, in relation to the same subject:

"At the end of the year, 1832, and in the beginning of 1833, M. Le Baron de Schilling constructed, at St. Petersburg, an electric telegraph, which consisted in a certain number of platinum wires, insulated and united in a cord of silk, which put in action, by the aid of a species of key, 36 magnetic needles, each of which were placed vertically in the centre of a multiplier. M. de Schilling was the first who adapted to this kind of apparatus, an ingenious mechanism, suitable for sounding an alarm, which, when the needle turned at the beginning of the correspondence, was set in play by the fall of a little ball of lead, which the magnetic needle caused to fall. This telegraph of M. de Schilling, was received with approbation by the Emperor, who desired it established on a larger scale, but the death of the inventor postponed the enterprise indefinitely."

Dr. Steinheil in his article "upon telegraphic communication," published in the London Annals of Electricity, states, "that the experiments instituted by Schilling, by the deflection of a single needle, seems much better contrived, than the arrangement which Davy has proposed, in which illuminated letters are shown by the removal of screens placed in front of them."

It would appear, that the French report is either incorrect, or that M. de Schilling had two plans in contemplation. His plan as intimated in the first and third extracts, is that of using a single needle in the form of a galvanometer, by means of which he made his signals, for instance, one deflection to the right might denote *e*; two *i*; three *h*: one deflection to the left *t*; two *s*; three *v*. His code of signals would then be devised in this manner:

rl	A	rrrl	K	llr	U
rrr	B	lrrr	L	lll	V
rll	C	lrl	M	rlrl	W
rrl	D	lr	N	lrlr	X
r	E	rlr	O	rllr	Y
rrrr	F	llrr	P	rlrr	Z
llll	G	lllr	Q	rrlr	&
rlll	H	lrr	R	lrrl	go on
rr	I	ll	S	lrll	stop
rrll	J	l	T	llrl	finish
rlrlr	1	lrlrl	6		

rrlrr	2	rrllr	7
rlllr	3	rllrr	8
lrrrl	4	llrll	9
lrrll	5	llrrl	0

If, however, his plan was that ascribed to him, by the Academy of Industry, of using 36 needles and 72 wires, it was exceedingly complicated and expensive, and was similar to that invented by Mr. Alexander, with the exception that Schilling used twice the number of wires.

[27] The Electro Magnetic Telegraph, of Counsellor Gauss and Professor William Weber, invented at Göttingen, 1833.

The deflection of the magnetic bar, by means of the multiplier, through the agency of the galvanic fluid, excited by the magneto electric machine, is the basis of their plan.

Fig. 51.

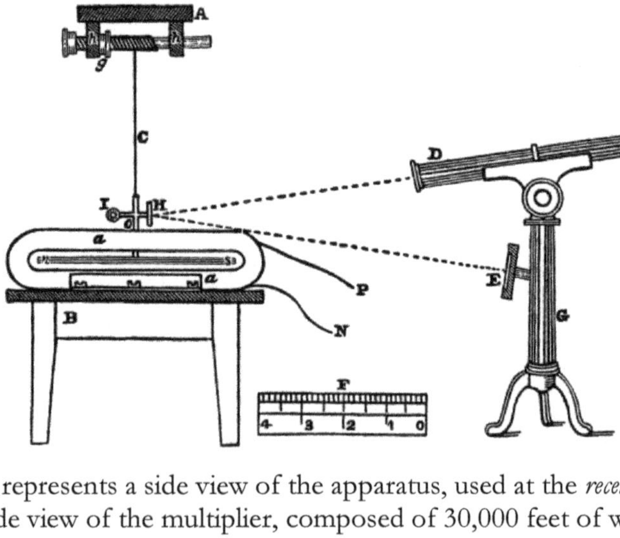

Figure 51 represents a side view of the apparatus, used at the *receiving* station: *a, a* is a side view of the multiplier, composed of 30,000 feet of wire, (almost 5½ miles,) upon a table, B: *n, s* is the magnetic bar, weighing 30 pounds, from which rises a vertical stem, *o*, upon which is a rod at right angles, supporting a mirror, H, on one end, and at the other a metallic ball, I, as a counteracting weight to that of the mirror. The magnetic bar is suspended by a small wire, fastened to the vertical stem, and at the top is wound round the spiral of the screw, *i*, which turns in the standards, *h'* and *h*, upon the platform, A, and which is secured to the ceiling. In the standards, *h'*, there is cut a female screw, of the same gradation as that upon which the wire is wound. By this means, the magnetic bar may be raised or let down, by turning the screw, without taking the bar from its central position in the multiplier: *g* is a screw for fastening the spiral shaft, when properly adjusted. P and N are the two ends of the wire of the multiplier. G is a stand for supporting the spy-glass, D, and also the case, E, into which slides the scale, F. The mirror, H, is at right angles with the magnetic bar, and presents its face to the spy-glass, D, as also to the scale at E. It is so adjusted, that the reflection of the scale at E, from the mirror, may be distinctly seen by the spy-glass. If the magnetic bar turns either to the right or left, the mirror must move with it, and if a person is observing it through the spy-glass, the scale will appear to move at the same time, thereby presenting to the eye of the observer another part of the scale than that seen when the bar is not deflected. The figures on the scale will show in what direction the bar has turned, and thus render it distinct to the observer, the only apparent object of the mirror, spy-glass and scale.

For the purpose of generating the galvanic fluid, they use the magneto electric machine. Their plan, being unwieldy and difficult to operate, is omitted, and in its stead, we introduce that form of it, invented by Dr. Page, which has already been described in figures 45, 46 and 47. There is also required for the purpose of making the desired deflections of the magnetic bar, a commutator, or pole changer, such as we have described in figures 48, 49 and 50. Figure 51 represents that portion of the apparatus at the *receiving* station. The magneto electric machine, and the pole changer, properly connected, are the instruments of the *transmitting* station. Two wires, or one wire and the ground, form the circuit between these two stations. The machine is put in operation by turning the crank, and the person sending the intelligence is stationed at the commutator, and directs the current through the extended wires to the multiplier of the receiving station, so as to deflect the bar to the right or left, in any succession he may choose, or suspend its action for any length of time.

"But in the apparatus for observation, the observer looks into the spy-glass, and writes up the kind and results of the variations of the magnetic needle. In order to have a control of the recorder, let there be a good number of spy-glasses directed towards the same mirror, in which observers may watch independently of each other. Suppose that five variations of the magnetic needle signifies a letter. L denotes a variation to the left, and R to the right. Then, might r r r r r denote A; r r r r l denote B; r r r l r denote C; r r l r r denote D; and so on. In the whole, we obtain, by the different arrangements of the five, which are made with the two letters, R and L, 32 different telegraphic signs, which may answer for letters and numbers, and of which we can select those where the most changes are introduced between *r* and *l*, as the most common letters, in order, in the best possible manner, to notice the constant variations of the magnetic needle."

The following would be the alphabetical signs, as arranged from the above directions:

A	rrrrr	I or Y	llrll	R	rrrll
B	rrrrl	K	lrrrl	S or Z	rrlrl
C	rrrlr	L	rlrrr	T	llrlr
D	rrlrr	M	rrlll	U	rlllr
E	rlrlr	N	lllll	V	lrrll
F	lrrrr	O	lrlll	W	llllr
G or J	lrlrr	P	lrlrl		
H	rlrrl	Q	llrrr		

Numerals.

1	rllll	6	rllrr
2	rrllr	7	lllrl
3	rlrll	8	llrrl
4	rllrl	9	lrrlr
5	lllrr	0	lrllr

It will be seen, that, by representing the letters and numerals with these variously combined deflections of the needle, words and sentences may be transmitted. At the end of each letter there is a suspension of the action of the bar for a short time, and at the end of a word, a still longer pause. This plan of an electric telegraph was tried for a distance of one mile and a quarter, in Göttingen. Of its further success, we are not informed.

Experiment of Messrs. Taquin & Ettieyhausen.[28]

"Messrs. Taquin and Ettieyhausen made experiments with a telegraphic line over two streets in Vienna, 1836. The wires passed through the air and under the ground of the Botanic garden."

No other account appears to have been given of their experiments than that quoted above.

Electro Magnetic Printing Telegraph, invented by Alfred Vail, September, 1837.

Soon after my connection with Professor Morse as copartner, and at the time I was constructing an instrument for exhibiting the advantages of his telegraph to a committee of Congress, it occurred to me, that a plan might be devised, by means of which the letters of the alphabet could be employed in recording telegraphic messages. I immediately gave it my attention, and produced the following plan:

Figure 52 represents a front and side view of the instrument.

Figure 55 is a top view.

Figure 56 is a back view.

The same parts are represented by the same letters in the three views. In figure 52, Q, Q is the platform upon which the whole instrument is placed. M and M are wooden blocks supporting parts of the instrument, K is the helix of the soft iron bar, H, passing through its centre, and there is another coil and bar directly behind this; the two making the electro magnet. G is its armature, fastened to the lever, F, F, which has its axis at I, (seen in figure 55, at X, X.) R is a brass standard for supporting the lever, F, upon its axis, by means of two pivot screws: *a* and *a* are two screws passing vertically, through the standard, R, for limiting the motion of the lever, F, F. J is a spiral spring, at its upper end, fastened to the lever, F, and at its lower end passes through the screw, L, by which it is adjusted, so as to withdraw the armature from the magnet, after it has ceased to attract, and for other purposes, hereafter to be explained. N and O is a brass frame, containing the type wheel, B', and the pulley, E and U. P and P represent the edge of a narrow strip of paper, passing between the type wheel and pulley, E. D is the printer, which, at the bottom, forms a joint with the end of the lever F and *r*. B represents twenty-four metallic pins, or springs, projecting at right angles from the side of the type wheel; each pin corresponding in its distance from the centre of the type wheel, to its respective hole, represented by dots upon, the index, C; so that if the pin is put in any one of the holes, the type wheel, in its revolution, will bring its corresponding pin in contact with it.

Fig. 52.

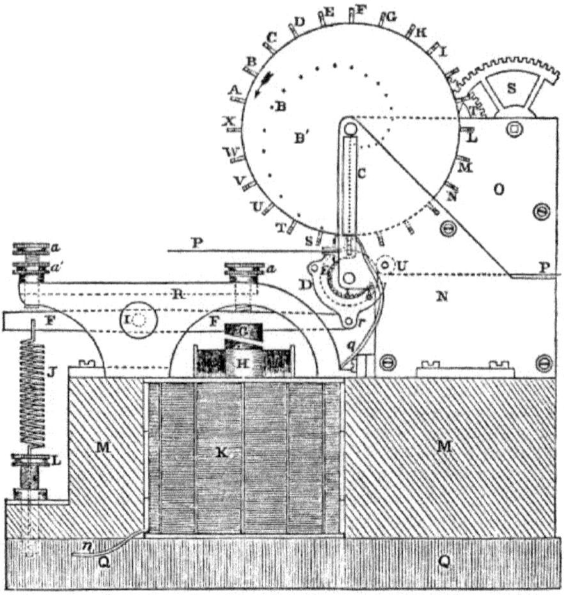

There are 24 holes corresponding to the following letters of the alphabet. A, B, C, D, E, F, G, H, I, K, L, M, N, O, P, Q, R, S, T, U, V, W, X, and the types are lettered accordingly. The cog wheels, T and S, are a part of the train of the clock. The lever, F, F, has two motions, one up and another down, and both are employed by an attachment at the end of the lever, *r*, and in the following manner: figures 53 and 54 represent a front and end view of the roller, E, and printer, D, (figure 52,) enlarged. D is the printer, figure 53, of the form shown by D, (figure 54.) E is the roller over which the paper, P, is carried. A is the front of the type having ears, *h, h*, projecting from each side. Through the sides of the printer, D, D, a rod, U, passes, in order to give more firmness to the frame. The rod projects a little on each side of the frame at J, J. These projections slide in a long groove in the frames, N and O, figure 52, by which the printer is kept in its position, and allowed freely to move up and down. It will be observed that the upper parts of the frame, D, D, extends over the top of the roller, E, and nearly touch each other, but are so far separated, as to let the type, A, of the type wheel, in its revolution, freely pass between them: *d', d'*, are the sides of the joint, which are connected with the lever, F, fig. 52. From the construction of this part, it will appear that if the printer, D, is brought down by the action of the magnet upon the lever, the two projections, *k, k*, will come in contact with the ears, *h, h*, and bring the type in contact with the paper upon the roller, E, and produce an impression. In figure 54 is shown a ratchet wheel, *i*, on the end of the roller, E, a catch, *e*, and spring, *c'*, adapted to the ratchet. Upon the release of the lever, F, fig. 52, the spring, J, will carry down the lever on that side of its axis, and up at *r*, which will cause the roller, E, to turn, and consequently the paper, P, to advance so much by the action of the catch, *e*, upon the ratchet wheel, as will be sufficient for printing the next letter.

Fig. 53. Fig. 54.

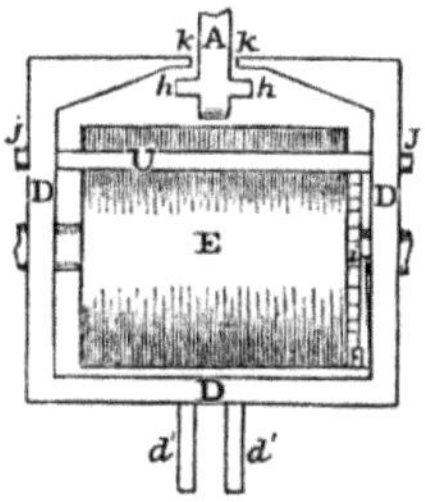

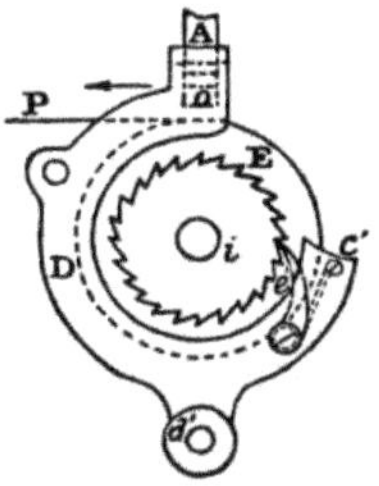

Figure 55 represents a top view of the machine. S is the barrel upon which is wound a cord, sustaining a weight which drives the clock train, and upon the same shaft with it is a cog wheel driving the pinion, *m*, on the shaft, T; and on the same shaft, T, is another cog wheel, driving the pinion, *n*, of the type wheel shaft, I'. K and K, are the helices of the large magnet, of which H and H are the soft iron arms. M, M, M, M, are the blocks which support the instrument. F and F is the lever, *a* and *a* its adjusting screws; x' and x' its axis; *k* and *k* are the two upper coils of the two electro magnets at the back part of the instrument for purposes hereafter to be described; *x* is the wire soldered to the plate buried in the ground; *p* is the wire proceeding to the battery; *c* is the connecting wire of the two electro magnets, *k* and *k*; w is the support of the pendulum; *v* is the escapement wheel; A is the type wheel; D and D is the printer, and B the roller over which the paper, P, is carried.

Fig. 55.

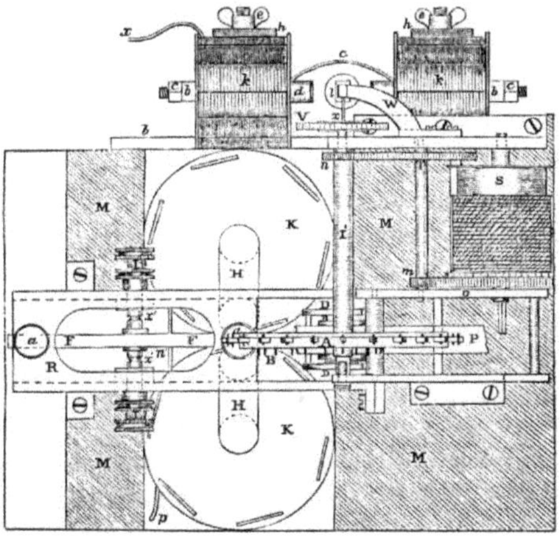

Fig. 56.

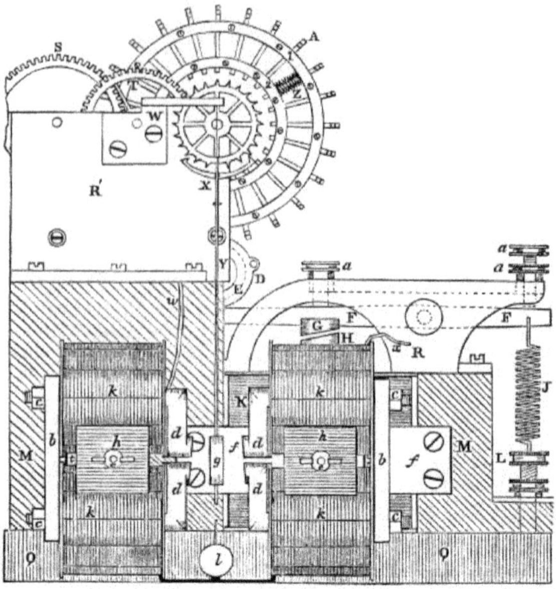

Figure 56 represents a back view of the instrument; *k*, *k* and *k*, *k* are the coils of two electro magnets, surrounding the soft iron bars, *d*, *d* and *d*, *d*; *b* and *b* are the flat bars through which *d*, *d* and *d*, *d* pass, and are fastened together by the screw nuts *c*, *c* and *c*, *c*. The right hand electro magnet is fastened to the blocks, M and M, by the support, *f* and *f*; from which proceeds a bolt passing between the coils, *k* and *k*, and the block, *h*, with a thumb-nut upon it, by which the whole is permanently secured. In the same manner the left hand magnet is secured to the block, M. R' is the outside portion of the brass frame containing the clock work. W is a standard fastened to R', for supporting the pendulum, Y. X, Y, and *l* are parts common to a chronometer for measuring the time, viz. the escapement and pendulum. The escapement wheel has 24 teeth, corresponding in number with the type on the wheel, and such is the arrangement of the parts, that when the pendulum is upon the point of return, either on the right or left hand, a type is directly over the paper, and the armature, *g*, is near the face of one or the other of the magnets; so that, if an impression is to be made with the type, thus brought to the paper, the pendulum, Y, is ready to be held by the magnet at the same time from making another swing until the type has performed its office, which will be hereafter explained.

A shows the type as they are arranged on the wheel. The types are square, and move freely in a groove, cut out of the brass type wheel. At 1 and 2 are seen flat brass rings, which are screwed to the wheel, and over the types, confining them to their proper places. Z is a spiral spring, of which there is one to each type, by means of which the type is brought back to its former position, after it is released by the printer. Through each type there is a pin, against which the inner end of the spiral spring rests. The outer end of the spring rests against the circular plate. W represents the wire from the upper helix, soldered to the metallic frame, R'. The two helices of the left hand magnet are joined together, and from the bottom helix the wire proceeds to the lower coil of the right hand magnet. These two helices are likewise connected, and the wire leaves the upper coil at *x*. Thus the wire is continuous from *w* to *x*. From *x*, the wire is continued to a copper plate, buried in the earth. The frame, R', being brass, the arbor of the type wheel, and the wheel itself, and each being in metallic contact, they answer as a continuous conductor with the wire, *w*, for the galvanic fluid.

The index, *c*, figure 52, is insulated from the frame, N, being made of ivory. There is inserted in the ivory, a metal plate, containing the holes, to which is soldered a wire, *q*, connected with the back coil, K. The two helices being connected, the wire of the front helix comes off at *p*, and from thence is connected with one pole of the battery; from the other pole, it is extended to the distant station, and is there connected with a similar instrument. It will be observed, that the circuit is continuous, except between the type wheel and the metal plate in the ivory. When neither station is at work, the batteries of both are thrown out, and their circuits, retaining in them the magnets of both stations, are closed. For this purpose, there is an instrument at each station, resembling in some respects the pole changer, figures 48, 49 and 50. If one of the stations wish to transmit by reversing his circuit instruments, the battery is instantly brought into the circuit. Through the agency of the clock work and weight, and the pendulum, both instruments are vibrating together, and their type wheels are so adjusted, that when *A type*, of one station, is vertical, the *A type*, of the other station, is also vertical. Now, suppose one station wishes to transmit to the other, the word *Boston*, for example: he first brings his battery in the circuit, then places a metallic pin in the hole of his index, C, marked for the letter B. When the type wheel shall have brought round the pin, corresponding to the type, B, on the wheel, its pin will come in contact with the inserted pin of the index, and instantly the circuit is established. The fluid, passing through the coils of the magnets, on each side of the pendulum, will hold it, and also passing through the coils, K, will bring down the lever, F, F, and with it, the printer, D, which, as heretofore described, in figures 53 and 54, will bring the type, with considerable force, against the paper. The instant the two pins have come in contact with the moving pin, it is taken out and put in the hole, O, when the same operation is performed, and in like manner for the remaining letters of the word. The pin can be so arranged, as to be thrown out the instant a complete contact is made.

The rapidity of this printing process would be as follows: Suppose the pendulum makes two vibrations in a second; that is, it goes from right to left in half a second, and returns in half a second. Since, then, a single letter is brought to the *vertical position*, ready to be used if needed, at the end of each vibration, it is clear, that two letters are brought to the vertical position every second, or 120, every minute. This is not, however, the actual rate of printing; for, in the word *Boston*, the type wheel, after B is printed upon the paper, must make so much of a revolution as will bring the letter O to the paper. This will require 12 vibrations of the pendulum; S will require 4; T, 1; O, 18, and N, 22; equal to 57, to which add 6, the time required to print each letter, will make it 63. This, divided by 2, gives 31½ seconds, the time necessary to print 6 letters. If we now take an ordinary sentence, and estimate, in the same manner, the time required to print it at the distant station, we shall be able to find what number of letters it can print per minute.

"There will be a declaration of war in a few days, by this government, against the United States. Orders have just been received to have all the public archives removed to Jalapa, which is sixty miles in the interior, for safe keeping."

Here are 184 letters, and would require 2266 vibrations, to which add 184, the number of letters would give 2450 half seconds, equal to 1225 seconds, the time required for printing the message; or over 20 minutes; the rate being six and two-thirds seconds for each letter.

If, however, a vocabulary is used, with the words numbered, and instead of using the 26 letters of the alphabet on the type wheel, we substitute the 10 numerals, in their place, we reduce the time required for a revolution of the wheel, and it is clear that this same message may be transmitted in much less time.

The following numbers represent the words of the same message, in the numbered vocabulary: 48687, 54717, 4165, 1, 12185, 34162, 54078, 25393, 1, 18952, 11934, 6177, 48766, 21950, 1106, 48652, 51779, 46532, 34475, 22991, 28536, 4321, 40254, 49085, 22991, 1391, 48652, 39087, 3845, 41278, 49085, 28536, 54536, 28668, 45008, 31634, 25393, 48652, 27326, 19865, 42813, 28592. Here are 42 numbers, and 196 figures. To 196 add 42, the spaces required, and we have 238 impressions to make, to write the sentence thus represented. By calculation, we find there is required, in order to bring each numeral and space in its proper succession, to the vertical position, 1624 vibrations of the pendulum, which, at the rate of two to the second, gives the time required to transmit the message at 812 seconds, or nearly 13 minutes, being at the rate of 18⅓ letters per minute.[29]

If, however, the vibrations of the pendulum are increased at the rate of 4 in a second, then the time required for the transmission of the message would be almost 7 minutes, and at the rate of 36⅔ letters per minute.[30] If it be increased to 6 vibrations per second, then the time would be 4½ minutes, and at the rate of 55 impressions per minute.

The modes of using the English letter for recording telegraphic messages are various, and they may be classed, as, First, Those which are rapid in transmission; expensive in construction, and complicated in machinery. Second, The less rapid in transmission; economical in construction, and simple in its machinery. Third, The slow in transmission; less expensive than the first class in construction; but complicated in its machinery.

To the *first* class, belong those using 26 types; one for each of the letters of the alphabet, and 13 extended wires, from station to station, with more or less battery. These types are arranged in a row, directly over the paper which receives the impression, and consequently require a strip of paper some 4 or 5 inches broad. Each type is furnished with an electro magnet and lever, answering as a hammer to bring down the types upon the paper. As the types are arranged in a straight line, they would present the following order:

A	B	C	D	E	F	G	H	I	J	K	L	M	N	O	P	Q	R	S	T	U	V	W	X	Y	Z
-				-		-	-	-			-		-		P		R								
-				-		-	-	I			-		N		-		-		T						
-				-		-	-	I			-		N		-		-		-						
-				-		G	-				-				-		-		T						
-				E		-	-				L				-		-								
-				E		G	-								-		R								
A						-	-								P										
							H																		

Here we have the style of this kind of printing. By spelling the letters on the first line, then on the second, and so on, the words "Printing Telegraph" can be made out. Those letters which follow each other in the word, and also follow each other in the alphabet, are placed upon the same line, but when a letter occurs preceding the last, a new line must be taken, otherwise the word cannot be read. It will appear, that in this mode, sometimes two or three, or four letters, may be printed at one and the same instant, where they succeed each other in alphabetical order. This plan is extremely rapid for *one instrument*, but extremely slow for *thirteen wires*.

Supposing two such instruments are used upon a line of 40 miles, and suppose the wire to cost per mile, fifty dollars. The expense for wire alone would be $26,000. There are other expenses which we will omit in this, as well as those plans which will be described hereafter. Let it be assumed, in order to make equal comparison throughout, that the number of successive motions of the type lever, in these various plans about to be given, are 4 to a second. But as this instrument may make, with two or more of its levers, two or more impressions per minute, let it be 8 instead of 4 per second. It will then be capable of transmitting 480 letters per minute. With all this, there are many disadvantages, which will be developed as we proceed.

Under the same class, there is another plan, using the 26 types upon the ends of as many levers, each lever employing the electro magnet, and the line consisting of 13 wires. In this arrangement the types are made to strike in any succession required by the message, at the *same point* upon the paper, *falling back* and resuming their first position, after having printed their letter, in order to allow the next type to occupy the same point previously occupied by the other. The printing of this plan will appear on paper as ordinary printing. Thus, PRINTING TELEGRAPH. If we suppose that 4 hammers, carrying type, can strike the *same point* in a second, and each resume their original position in succession, thus passing each other without collision, it may print at the rate of 240 letters per minute.[31] The instrument would be a complicated one and subject to derangement.

To the *second* class, belong all those which print in letters of an hieroglyphical character. The *first* plan is that employing one wire and one motion. Under this head, is that of Prof. Morse's. He employs but one wire and one electro magnet for printing, which has but one motion. Suppose this to be capable of operating with the same speed as the preceding, viz. four motions per second. The telegraphic alphabet as adopted by Prof. Morse require for each letter the following number of motions of the type or pen lever, as lines require time in proportion to their length, they are so estimated: A 3, B 5, C 4, D 4, E 1, F 4, G 5, H 4, I 2, J 6, K 5, L 5, M 4, N 3, O 3, P 5, Q 5, R 4, S 3, T 2, U 4, V 5, W 5, X 5, Y 5, Z 5.

If we take the *standard number* of types for each letter constituting it printer's case, considering Z as 2, we shall have A 85, B 16, C 30, D 44, E 120, F 25, G 17, H 64, I 80, J 4, K 8, L 40, M 30, N 80, O 80, P 17, Q 5, R 62, S 80, T 90, U 34, V 12, W 20, X 4, Y 20, Z 2. The whole number of letters are 1177. The number of motions required to transmit them would be 3420, to which add, one motion for the time required to space a single letter, and we have 4597 motions, made in printing 1177 letters which will make the average number of motions to each letter $3\,{}^{1066}/_{1177}$, nearly 4. Let it be 60 per minute. Expense for one wire of 40 miles, $2000.

Second plan, is that where two wires are used, two magnets, two type levers, and the telegraphic characters, such as are represented in table 1, page 30. The first three letters require three motions each; the next 16, require 2 each, and the last 7, require 3 each. Taking the 1177 letters, the motions required to transmit them in the characters of this alphabet, would be, 2195 + 1177 for spaces and would equal 3372, which divided by 1177, would give the average number of motions at $2\,{}^{1018}/_{1177}$ for each letter, nearly three or 80 per minute. Cost of wire $4000.

Third plan, is that using three wires, three magnets, three type levers and the telegraphic characters represented in table second, page 30. The seven first would require one motion each, and the remainder two each. Taking 1177 letters, the motions required to transmit them, would be 1917 + 1177 for spaces, and would equal 3094 motions, which, divided by 1177, would give the average number of motions $2^{740}/_{1177}$ for each letter, nearly 2⅔, or 85 letters per minute. Cost of wire $6000.

Fourth plan consists in using four wires, four electro magnets, four type levers, and the telegraphic characters of the third table. The first sixteen letters require the time of but one motion each; the remainder, two each. Using 1177 letters, the motions required to transmit them would be 1506 + 1177 for spaces, and would equal 2683, which divided by 1177, would give the average number of motions $2^{329}/_{1177}$ for each letter, nearly 2⅓, or 103 letters per minute. Cost of wire $8000.

Fifth plan, is that of using five wires, five electro magnets, five type levers, and the telegraphic characters of the 4th table. The characters would require one motion each, equal to 1177 + 1177 for spaces, and would equal 2354, which, divided by 1177, would give the average number of motions, 2 for each letter, or 120 letters per minute. Cost of wire $10,000.

We now come to the *third* class, in which 26 types are used, arranged upon the periphery of a wheel, in alphabetical order, and require to be brought to one certain point, where the paper is ready to receive the impression of the type, by another arrangement, distinct from the type wheel and its machinery. Of this plan, is that which has been already described in figures 52, 55 and 56. The estimate is there carried out, at 4 motions per second, gives 36⅔ letters per minute. Cost of wire $2000.

The following table will show the comparative value of these various methods:

		Letters per minute.	**Cost.**	**Number of wires.**	**On Morse's plan.**	**No.**
1st Class.	1st plan,	480	$26,000	13	780	1
	2d "	240	26,000	13	780	2
2d Class.	1st plan,	60	2,000	1	60	3
	2d "	80	4,000	2	120	4
	3d "	85	6,000	3	180	5

		Letters per minute.	Cost.	Number of wires.	On Morse's plan.	No.
	4th "	103	8,000	4	240	6
	5th "	120	10,000	5	300	7
3d Class.	1st plan,	37	2,000	1	60	8

We find by comparison that Morse's plan, No. 3, of using a single wire, with a single instrument, produces 60 characters per minute; while No. 1, with 13 wires, and one instrument, produces 480 characters per minute. Let, however, the 13 wires be multiplied by 60, (the number of characters which a single instrument of the plan, No. 3, can transmit,) the number of characters which 13 wires, with 13 instruments would then produce, are 780 or 300 more than the *single instrument*, with 13 *wires*. The same comparisons may be made with the other plans, and it will be found that no advantage can be gained by their adoption.

All electro magnetic telegraphs require as their basis, the adoption of the *electro magnet*, where recording the intelligence is an object, and it would seem, must be applied in a manner equivalent to that mode adopted by Prof. Morse; that is, the application of the armature to a lever, and its single movement produced by closing and breaking the circuit. It is, therefore, safe to assume, that whatever improvement in one plan may be made to increase the rapidity of the movements of those parts of the telegraph which belong to the electro magnet, are equally applicable to any other plan, provided too much complication, already existing, does not counteract and defeat the improvement.

Some plans, however, use an extra agent besides the electro magnet, which is employed for measuring the time of the revolution of the type wheel, and the electro magnet is only called in, occasionally, to make the impression. In such plans the rapidity of communication demands the combined action, alternately, of both magnets. This, of course, increases the complication, and must certainly be considered a departure from other more simple arrangements. Whatever will reduce the inertia of mechanical movements and bring them to act with an approximate velocity, at least of the fluid itself, will increase the rapidity of transmission. The more the instrument is encumbered with the sluggish movements of material bodies, the less rapid, inevitably, must be its operation, even where several co-operating agents are assisting, in their respective spheres, to increase the rapidity of the motion.

Such is the case with the several kinds of letter printing telegraphs: very weighty bodies, comparatively speaking, are set in motion, stopped, again set in motion, and along with this irregular motion, other parts perform their functions. There must be a courtesy observed among themselves, or matters do not move on as harmoniously as could be desired. This is not always the case, especially where time is the great question at issue.

All printing telegraphs which use type, arranged upon the periphery of a wheel, must have, of necessity, these several movements, viz. the irregular revolution of the type wheel, stopping and starting at every division or letter; the movement of the machinery, called the printer; the irregular movement of the paper, at intervals, to accommodate itself to the letter to be printed; the movement of the inking apparatus, or what is not an improvement in cleanliness, paper of the character used by the manifold letter writer. So many moving parts, are so many impeding causes to increased rapidity, and are, to all intents and purposes, a *complication.*

The requirements of a perfect instrument are: economy of construction, simplicity of arrangement, and mechanical movements, and rapidity of transmission. To use one wire is to reduce it to the lowest, possible economy. If there is but one movement, and that has all the advantages which accuracy of construction, simplicity of arrangement and lightness, can bestow upon it, we might justly infer that it appeared reduced to its simplest form.

The instrument employed by Professor Morse has but a single movement, and that motion of a vibratory character; is light and susceptible of the most delicate structure, by which rapidity is insured; the paper is continuous in its movement, and requires no aid from the magnet to carry it.

The only object that can be obtained by using the English letters, instead of the telegraphic letters, is, that the one is in common use, the other is not. The one is as easily read as the other, the advantage then is fanciful and is only to be indulged in at the expense of time, and complication of machinery, increasing the expense, and producing their inevitable accompaniments, liability of derangement, care of attendance, and loss of time.

Wheatstone's Electric Needle Telegraph, invented in 1837.

The following description is taken from a pamphlet, published by T. S. Hodson, 15 Cross street, Hallon Garden, London, 1839, for the proprietors. It is unnecessary to copy the legal and technical wordy mass of the specification, embracing fifty-nine pages of closely printed matter of octavo size. A full description will be given, with the accompanying figures, so as to enable the reader fully to comprehend Mr. Wheatstone's plan.

His arrangement requires the service of five galvanometers, in every respect similarly constructed as that described by the figures 27, 28 and 29. Figure 57 is a representation of his dial, which is also a covering to the case containing, in the interior, the five galvanometers and their wires, (shown at the opening in the dial board,) and numbered, 1, 1; 2, 2; 3, 3; 4, 4, and 5, 5. The coils of the multipliers are secured with their needles to the case, having each exterior needle projecting beyond the dial, so as to be exposed to view. Of the wires from the coils, five are represented as passing out of the side of the case, on the left hand, and are numbered 1, 2, 3, 4 and 5. The other five wires pass out on the right hand, and are numbered in the same manner. The wires of the same number as the galvanometer, are those which belong to it, and are continuous. Thus the wire 1, on the left hand, proceeds to the first coil of galvanometer 1, then to the second coil, and then coming off, passes out of the case, and is numbered 1, on the right hand. So of the other wires, thus numbered. The dial has permanently marked upon it, at proper distances and angles, twenty of the letters of the alphabet, viz. A, B, D, E, F, G, H, I, K, L, M, N, O, P, R, S, T, V, W, Y. On the margin of the lower half of the dial are marked the numerals, 1, 2, 3, 4, 5, 6, 7, 8, 9 and 0. The letters C, J, Q, U, X, Z, are not represented on the dial, unless some six of those already there are made to sustain two characters each, of which the specification is silent. Each needle has two motions; one to the right, and the other to the left. For the designation of any of the *letters*, the deflection of two needles are required, but for the *numerals*, one needle only. The letter intended to be noted by the observer, is designated, in the operation of the telegraph, by the *joint deflection* of two needles, pointing by their convergence to the letter. For example, the needles, 1 and 4, cut each other, by the lines of their joint deflection, at the letter V, on the dial, which is the letter intended to be observed at the receiving station. In the same manner any other letter upon the dial may be selected for observation. Suppose the first needle to be vertical, as the needles 2, 3 and 5, then needle 4 being only deflected, points to the numeral 4, as the number designed.

Fig. 57.

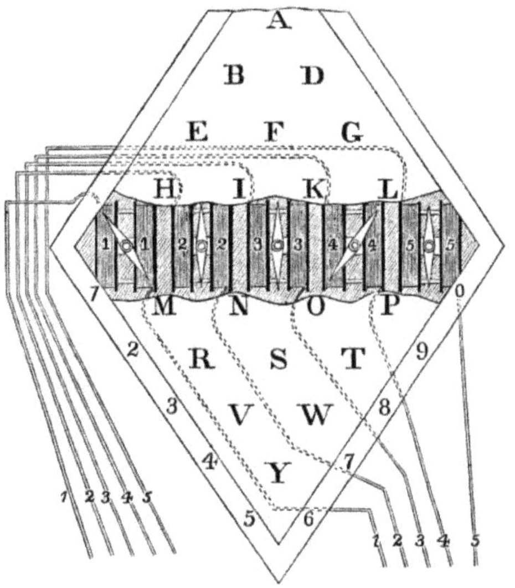

We will now proceed to describe the arrangement of the springs and buttons upon the platform, C, C, figure 58, (representing a top view,) by the operation of which, any two needles may be deflected to designate a letter, or one needle to designate a numeral.

Fig. 58.

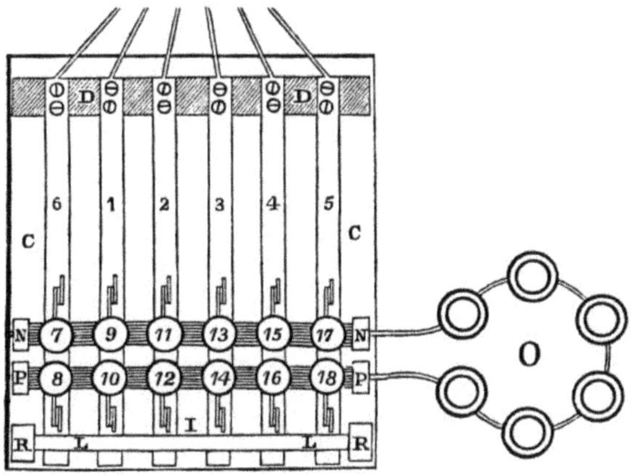

The numbers 6, 1, 2, 3, 4 and 5, represent keys of thin brass, and elastic, and are each fastened to a wooden support, D, D, by means of two screws. These keys are continued under and project beyond, the brass bar, L and L, which is supported by two standards, R and R. Whenever these keys are not pressed upon, they are each in *metallic contact* with the *bar*, R and R. The numbers 7, 8, 9, 10, &c. represent ivory buttons with a metallic stem beneath them, passing through a hole in the spring, or key, and on the lower side of the spring the stem is enlarged, so as to form a kind of hammer, designed to make a metallic contact with the two brass bars, beneath the springs, and represented as supported by the standards, N and N and P and P. Each of the buttons have a small wire spiral spring, to which they are fastened, and the small spring is itself fastened to the larger spring. O represents the galvanic battery, with its poles in connection with the two metallic bars, N and P.

Figure 59 represents a side view of the key arrangement. F is the platform. E the wooden support of the six keys. H is the larger spring, or key, secured to the support by screws, *h*. The spring is observed to project beyond the metallic cross bar, L, after passing beneath it. R is the support of the cross bar, L. N and O are two of the ivory buttons, upon their spiral springs, *a* and *c*. Below the button, O, is a shoulder, formed at *i*, upon the stem which passes through the spring, H, and another shoulder is formed by the hammer, *u*, below the spring. It will be observed, that two buttons of the same key are never used at the same time. If the button, O, is to be pressed down, the weaker spring, *c*, will permit it to descend until the upper shoulder comes in contact with the larger spring, H, when more pressure is applied, and that spring is brought down, breaking its contact with the metallic cross bar, L, until the hammer, *u*, comes in contact with the metallic plate, *n*, upon the support, K, and as the plate, *n*, is connected with N pole of the battery, the connection is formed with it. It will, however, be noticed, that the button, N, not being pressed upon, *will not*, (though it descends with the larger spring,) be brought in contact with the other plate upon the support, J, and connected with the positive pole of the battery. To the end of each spring, a wire, S, is soldered, the purpose of which will be shown hereafter.

Fig. 59.

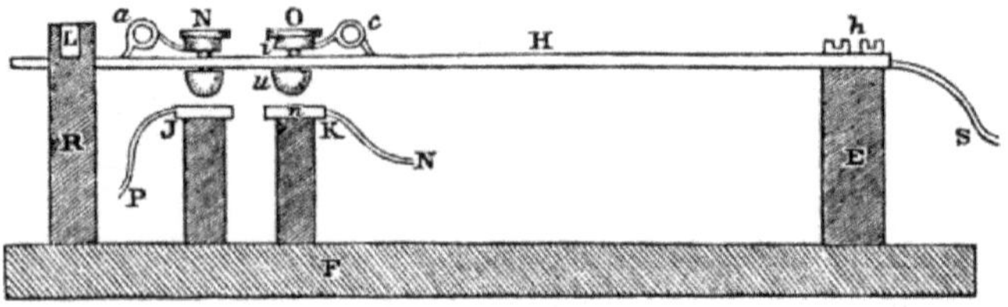

Fig. 60.

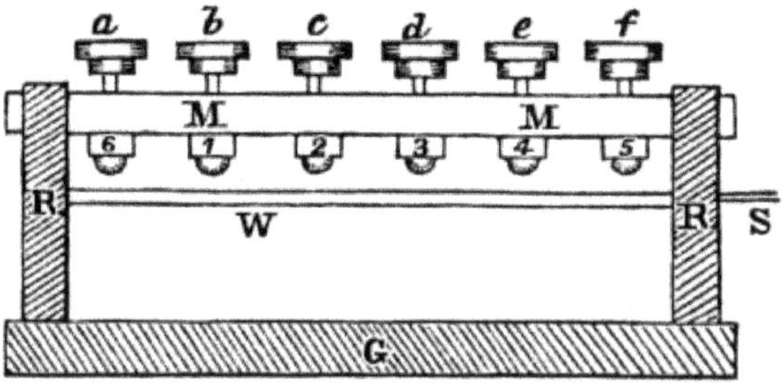

Figure 60 represents an end view of the key arrangement; *a*, *b*, *c*, *d*, *e*, *f*, are the buttons, M and M the metallic cross bar, beneath which are seen the ends of the six larger springs, 6, 1, 2, 3, 4 and 5. R and R are the supports of the bar, M and M. G is the platform. W is the support of the metallic plates, with which the hammers of the little keys, or buttons, come in contact. S the wire leading to the battery.

Having shown the several parts of Mr. Wheatstone's plan, we will proceed to describe the arrangement of two termini, as prepared for transmitting intelligence. Figure 61 represents the arrangement of one station, which we may suppose to be PADDINGTON. Figure 62 represents the plan of the other station, which we will suppose to be SLOUGH. The distance between these two places is eighteen miles.

In figure 61, it will be seen, that a wire is soldered to the end of each of the springs 6, 1, 2, 3, 4 and 5, and are respectively connected with the five wires of the dial, and the common communicating wire, number 6, which does not pass through the dial, nor is connected with any of the galvanometers. On the right hand side of the dial, the wires are extended until they are shown as broken. From this point to the opposite one, figure 62, where the wires appear also as interrupted, we may suppose 18 miles to intervene. The wires here proceed to the dial of the Slough station, making their proper connections with their respective galvanometers, and from thence are continued and soldered to their springs of the key arrangement, with the exception of wire, number 6, which passes direct to the key, 6, without going through the dial case. In both figures, is represented the battery, O, consisting of six cups. The wire from one pole of the battery is connected with the N metallic plate, the other wire with the P metallic plate. While none of the buttons are pressed down, the battery is *not* in action, and it will also be observed that the circuits are all *complete*. The action of the keys, then, is this, by a single operation to break the circuit formed with the cross bar, L, L, and, at the same time, bring *into* the circuit, the battery, O.

The following numbers, representing the buttons, are those necessary to be pressed down, in order to signal the letters and numerals on the dial:

Letters.

For	A,	buttons	10	and	17.	For	M,	buttons	9	and	12.
"	B,	"	10	"	15.	"	N,	"	11	"	14.
"	D,	"	12	"	17.	"	O,	"	13	"	16.
"	E,	"	10	"	13.	"	P,	"	15	"	18.
"	F,	"	12	"	15.	"	R,	"	9	"	14.
"	G,	"	14	"	17.	"	S,	"	11	"	16.
"	H,	"	10	"	11.	"	T,	"	13	"	18.
"	I,	"	12	"	13.	"	V,	"	9	"	16.
"	K,	"	14	"	15.	"	W,	"	11	"	18.
"	L,	"	16	"	17.	"	Y,	"	9	"	18.

Numerals.

For	1,	buttons	7	and	10.	For	6,	buttons	8	and	9.
"	2,	"	7	"	12.	"	7,	"	8	"	11.
"	3,	"	7	"	14.	"	8,	"	8	"	13.
"	4,	"	7	"	16.	"	9,	"	8	"	15.
"	5,	"	7	"	18.	"	0,	"	8	"	17.

Fig. 61.
PADDINGTON.

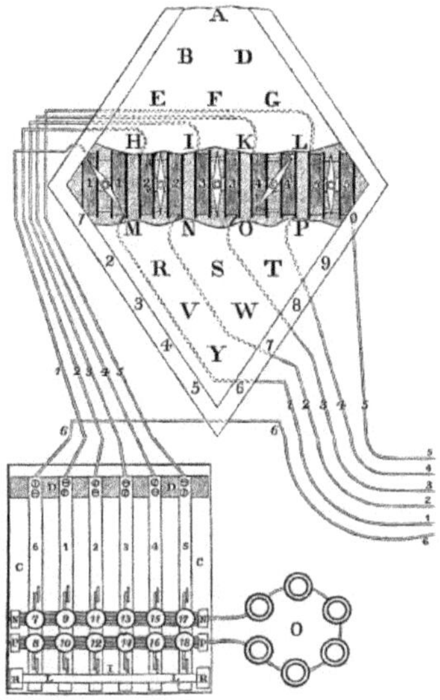

Fig. 62.
SLOUGH.

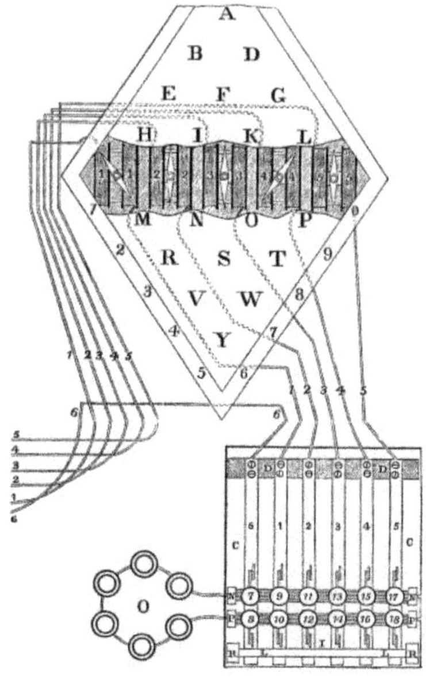

The direction of the current, when the letter V is to be signalled, is this: pressing down the buttons, 9 and 16, at the Paddington station, the fluid leaves the battery, O, along the wire to the cross bar, P; then to the hammer of the button, 16; then to the spring, 4; then along wire, 4, to the galvanometer, 4, and through it, deflecting the lower half of the needle to the left; then along the extended wire, 4, to the dial, and galvanometer, 4, of the Slough station, deflecting the lower half of that needle to the left; then to wire, 4, leaving the dial, to key, 4; then to the cross bar, L and L; and along the cross bar to key, 1; then to wire, 1; then to galvanometer, 1; and through it, deflecting the lower half of the needle to the right; thence it proceeds along the extended wire, 1, to the Paddington station; entering the dial to the galvanometer, 1, deflecting the lower half of the needle to the right; then along wire, 1, to the key, 1; then to button, 9; then to the cross bar, N, beneath; and then to the negative pole of the battery, O. It will be observed, that the needles of both stations, thus deflected, point to the same letter, V. In Mr. Wheatstone's arrangement, but one person can transmit at the same time, although he uses six extended wires. One must wait while the other is transmitting.

If a numeral is to be signalled, it is obvious, that but one galvanometer is needed. We will, therefore, suppose that the needle, 1, is vertical.

Let the buttons, 7 and 16, be pressed down, at the Paddington station. The current then leaves the positive pole of the battery, O, to the cross bar, P; then to the key, 4; then along wire, 4, to galvanometer, 4, deflecting the lower half of the needle to the left; from thence to the Slough station to galvanometer, 4, deflecting the lower half of the needle to the left; then to wire, 4; then to key, 4; then to the cross bar, L and L, and along it to key, 6; then to wire, 6, and along the extended wire to the Paddington station, to key, 6; then to the cross bar beneath the button, 7; then to the negative pole of the battery, O. The needles, 4 and 4, of both stations, are simultaneously deflected, so as to point to the figure, 4, on the margin of the dial.

In this manner the circuits required for each letter and numeral may be traced out. Now, suppose the message to be sent from the Paddington station to the Slough station, is this, "WE HAVE MET THE ENEMY AND THEY ARE OURS." The operator at Paddington presses down the buttons, 11 and 18, for signalizing upon the dial of the Slough station, the letter W. The operator there, who is supposed to be constantly on the watch, observes the two needles pointing at W. He writes it down, or calls it out aloud, to another, who records it, taking, according to a calculation given in a recent account, two seconds at least for each signal. Then the buttons, 10 and 13, are pressed down, and the needles are observed to point at E; and so for the remaining letters of the sentence, U excepted, which has no letter on the dial.

The peculiarity of Mr. Wheatstone's plan, is, the employment of six wires for one *independent* line of communication. The use of five galvanometers, with their needles, by the deflection of which, 30 letters and numerals are pointed out. The messages are not recorded by the instrument itself, but it is necessary that a person be constantly observing the successive movements of the needles, and note them down as they point to the signal. This plan was invented in 1837, and as Prof. Wheatstone took out letters-patent in the United States, in 1840, for this arrangement, it is a fair inference, that at that time, this was his simplest and most perfect method.

Steinheil's Electric Telegraph.

Description of the magneto electrical telegraph, erected between Munich and Bogenhausen, in 1837, by Dr. Steinheil,[32] Professor of Mathematics and Natural Philosophy at the University of Munich, taken from the Annals of Electricity, Magnetism and Chemistry, conducted by William Sturgeon, London, April, 1839.

Fig. 63.

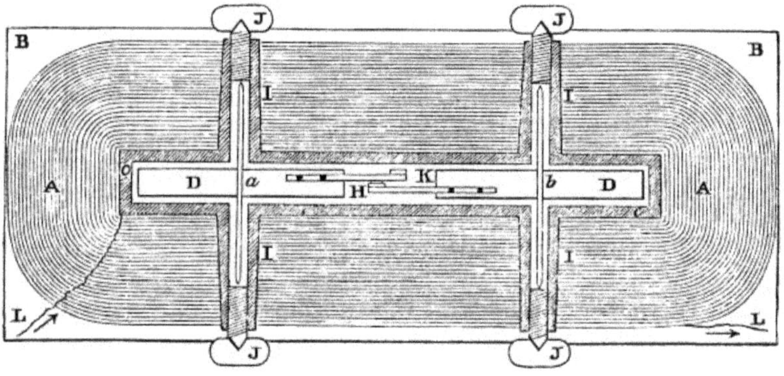

A, A represents a vertical section, through the centre of the coil of copper wire. C is the interior brass frame, round which the wire is wound. B and B are the sides of the frame; I, I, I, I are four brass tubes, soldered to the interior brass frame, and passing through the centre of the coil to its exterior, with a screw cut in the end of each; D and D are two permanent magnets movable on their axis, *a* and *b*. These spindles, *a* and *b*, on each side of the magnets, pass up the hollow of the tubes, and having their ends pointed, enter the centre cavity of the four thumb screws, J, J, J, J, by which they are supported, and delicately adjusted, so as to move easily and freely. L and L are the ends of the wire leaving the coil. H and K are two ink holders, attached to the magnets, which will be explained hereafter.

Fig. 64.

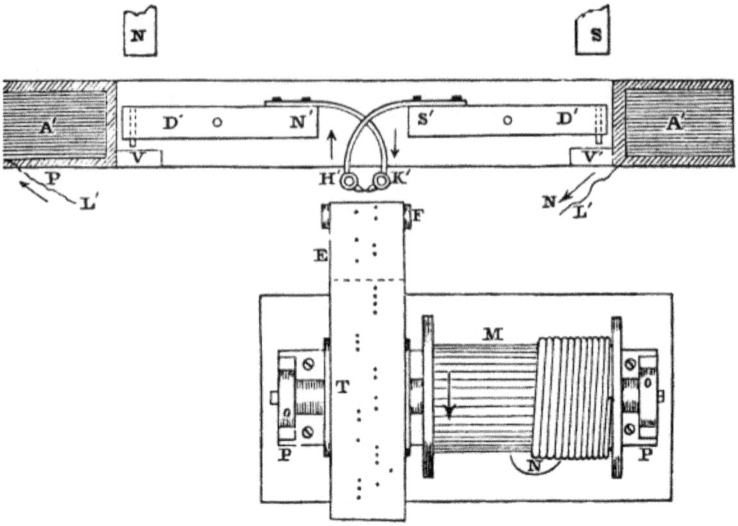

Figure 64 represents a horizontal section of the coil, and magnets D' and D', as above described, together with the other arrangements of the instrument for receiving intelligence. The magnetic bars are so situated in the frame of the multiplier, that the north pole, N', of the one, is presented to the south pole, S', of the other. To the ends which are thus presented to each other, but which, owing to the influence they mutually exert, cannot well be brought nearer, there are screwed on two slight brass arms, supporting little cups, H' and K'. These little cups, which are meant to be filled with printing ink, are provided with extremely fine perforated beaks, that are rounded off in front. When printing ink is put into them, it insinuates itself into the tube of their beaks, owing to capillary attraction; and without running out, forms at their apertures, a projection of a semiglobular shape. These little cups are seen at H' and K', and in figure 63 at H and K. The horizontal section shows, also, the position of the magnets in the instrument, with the beaks of the pens near the continuous band, or ribbon of paper, E, which is brought in front of the pens vertically from below, over a small roller, F. The paper is supplied from a large roll on a wooden cylinder, upon which is a cog wheel, and connected with a train of wheels and a vane, to regulate the rate of supply. The paper is drawn along before the pen by being wound upon a cylinder, T, concealed by the paper, and on the same shaft with the barrel, M, upon which is wound a cord supporting a weight, N, below. The shaft is supported in the standards, *o* and *o*, which are fastened to a plate of brass, P and P, also secured to the platform of the instrument. The barrel revolves in the direction of the arrow upon it.

When the electricity is transmitted through the coil of the indicator, both magnetic bars, D' and D', make an effort to turn in a similar direction upon their vertical axis, *a* and *b*. One of the cups of ink, therefore, advances towards the paper, while the other recedes. To limit this action, two plates, V and V', are fastened at the opposite ends of the free space, allowed for the play of the bars, and against which the other ends of the bars press. Only the end of one bar can, therefore, start out from within the multiplier at a time, the other being retained in its place. In order to bring the magnetic bars back to their original position, as soon as the deflection is completed, recourse is had to two small movable magnets, a portion of which is seen at N and S, whose distance and position are to be varied till they produce the desired effect. This position must be determined by experiment, inasmuch as it depends upon the intensity of the current called into play.

Having described the instrument, its operation is as follows: At the *transmitting* station is the pole changer, such as we have described in figures 48, 49 and 50, and the magneto electric machine such as is described in figures 45, 46, and 47, and are properly connected, and in the circuit with the instrument of the *receiving* station, such as we have just described. For one single circuit, one wire extends from the transmitting to the receiving station, the return half of the circuit is the earth. Thus the current passes from the generator along the extended wire to the receiving station, and to the copper plate, then returns through the ground to the copper plate of the transmitting station, to the pole changer and the magneto electric machine. Thus the circuit is complete.

It is clear, from what has preceded, that when the pole changer is thrown to the left side, (the machine being in operation,) the fluid is made to pass in the direction of the arrows, shown at P and N. Then the N' pole of the left hand magnet advances with its pen, K', to the paper, E, and a dot is made, and the S' pole of the right hand magnet recedes with its pen, H, from the paper, until the other end of the magnet strikes the stop, V'. Now, if the letter to be formed, requires two dots in succession from the same pen, the circuit is broken, and the fixed magnets, N and S, bring back the deflecting magnets, D' and D', to their former position, when the pole changer is again thrown to the left, and the magnets are deflected in the same manner as at first. Thus two dots are marked upon the paper, on the right hand line. But, now, let the pole changer be thrown to the right hand side, and the current is reversed. The N' pole of the left hand magnet, with its pen, K, recedes from the paper until it strikes the stop, V, and the S pole of the right hand magnet, with its pen, H', advances to the paper and makes its dot upon it on the *left* hand line. The pole changer is then instantly brought to the middle position, and the magnets resume their natural place, by the assistance of the stationary magnets, N and S. The sign which has been marked upon the paper during this operation is

· ·

·, and represents 9.

The following represents Mr. Steinheil's telegraphic alphabet:

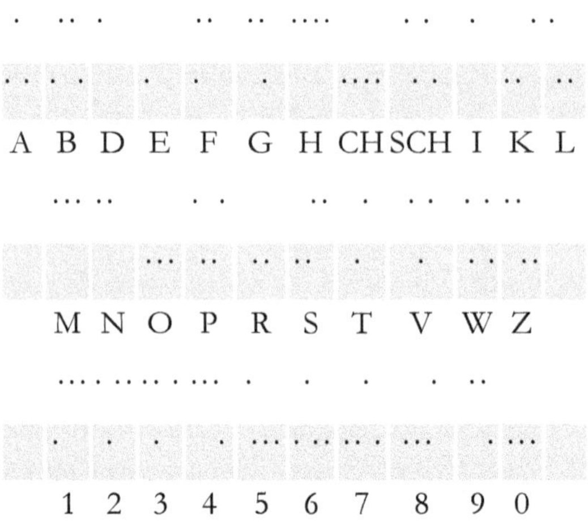

Masson's Electric Telegraph.

"In 1837, M. Masson, Professor of Philosophy at Caen, made trial of an electric telegraph, at the college of that city, for a distance of about 600 metres. He employed, for developing the galvanic current, an electro magnetic apparatus, similar, on the contrary, to that of Mr. Pixii, and made it act on magnetic needles placed at two ends of the circuit. Since that time, however, M. Masson has endeavoured to simplify and gradually improve his apparatus."[33]

Davy's Needle and Lamp Telegraph.

The following extracts from the London Mechanic's Magazine, vol. 28, page 296 and 327, 1837, is all the description we are able to find in relation to it:

"There is a case, which may serve as a desk to use in writing down the intelligence conveyed; and in this, there is an aperture about sixteen inches long, and three or four wide, facing the eyes, perfectly dark. On this the signals appear as luminous letters, or combinations of letters, with a neatness and rapidity almost magical. The field of view is so confined, that the signals can be easily caught and copied down without the necessity even of turning the head. Attention, in the first instance, is called by three strokes on a little bell; the termination of each word is indicated by a single stroke. There is not the slightest difficulty in decyphering what is intended to be communicated."

Extract from page 327.

"In front of the oblong trough, or box, described by your correspondent, a lamp is placed, and that side of the box next the lamp is of ground glass, through which the light is transmitted for the purpose of illuminating the letters. The oblong box is open at the top, but a plate of glass is interposed between the letters and the spectator, through which the latter reads off the letters as they are successively exposed to his view. At the opposite side of the room, a small key board is placed, (similar to that of a piano forte, but smaller,) furnished with twelve keys; eight of these have each three letters of the alphabet on their upper surfaces, marked A, B, C; D, E, F; and so on. By depressing these keys in various ways, the signals or letters are produced at the opposite desk, as previously described, how this is affected is not described by the inventor, as he *intimated* that the construction of certain parts of the apparatus *must remain* SECRET. By the side of the key board, there is placed a small galvanic battery, from which proceeds the wire, 25 yards in length, passing round the room. Along this wire the shock is passed, and operates upon that part of the apparatus which discloses the letters or signals. The shock is distributed as follows: The underside of the signal keys are each furnished with a small projecting piece of wire, which, on depressing the keys, is made to enter a small vessel, filled with mercury, placed under the outer ends of the row of keys; a shock is instantly communicated along the wire, and a letter, or signal, is as instantly disclosed in the oblong box. By attentively looking at the effect produced, it appeared as if a dark slide were withdrawn, thereby disclosing the illuminated letter. A slight vibration of the (apparent) slide, occasionally obscuring the letter, indicated a great delicacy of action in this part of the contrivance, and although not distinctly pointed out by the inventor, is to be accounted for in the following manner: when the two ends of the wire of the galvanic apparatus are brought together, over a compass needle, the position of the needle is immediately turned, at right angles, to its former position; and again, if the needle is placed with the north point southward, and the ends of the wire again brought over it, the needle is again forced round to a position at right angles to its original one. Thus, it would appear, that the slide or cover over the letters, is poised similarly to the common needle, and that by the depression of the keys, a shock is given in such a way as to cause a motion from right to left, and *vice versa*, disclosing those letters, immediately, under the needle so operated upon."

Alexander's Electric Telegraph, from the (Scotsmen) Mechanic's Magazine, Nov. 1837.

"A model to illustrate the nature and powers of this machine was exhibited on Wednesday evening at the Society of Arts in Edinburgh. The model consists of a wooden chest, about five feet long, three feet wide, three feet deep at the one end, and one foot at the other. The width and depth in this model are those which would probably be found suitable in a working machine, but it will be understood that the length in the machine may be a hundred or a thousand miles, and is limited to five feet in the model, merely for convenience. Thirty copper wires extend from end to end of the chest, and are kept apart from each other. At one end (which, for distinction's sake, we shall call the south end) they are fastened to a horizontal line of wooden keys, precisely similar to those of a piano forte; at the other, or north end, they terminate close to thirty small apertures, equally distributed in six rows of five each, over a screen of three feet square, which forms the end of the chest. Under these apertures on the outside, are painted, in black paint, upon a white ground, the twenty-six letters of the alphabet, with the necessary points, the colon, semicolon, and full point, and an asterisk, to denote the termination of a word. The letters occupy spaces about an inch square. The wooden keys, at the other end, have also the letters of the alphabet, painted on them in the usual order. The wires serve merely for communication, and we shall now describe the apparatus by which they work.

This consists, at the south end, of a pair of plates, zinc and copper, forming a galvanic trough, placed under the keys; and at the north end, of thirty steel magnets, about four inches long, placed close behind the letters painted on the screen. The magnets move horizontally on axes, and are poised within a flat ring of copper wire, formed of the ends of the communicating wires. On their north ends they carry small square bits of black paper, which project in front of the screen, and serve as opercula, or covers, to conceal the letters. When any wire is put in communication with the trough at the south end, the galvanic influence is instantly transmitted to the north end; and in accordance with the well known law, discovered by Oersted, the magnet at the end of that wire instantly turns round to the right or left, bearing with it the operculum of black paper, and unveiling a letter. When the key, A, for instance, is pressed down with the finger at the south end, the wire attached to it is immediately put in communication with the trough; and at the same instant, letter A, at the north end is unveiled, by the magnet turning to the right, and withdrawing the operculum. When the finger is removed from the key, it springs back to its place; the communication with the trough ceases; the magnet resumes its position, and the letter is again covered. Thus by pressing down with the finger, in succession, the keys corresponding to any word or name, we have the letters forming that word, or name, exhibited at the other end; the name VICTORIA, for instance, which was the maiden effort of the telegraph on Wednesday evening."

Fig. 65.

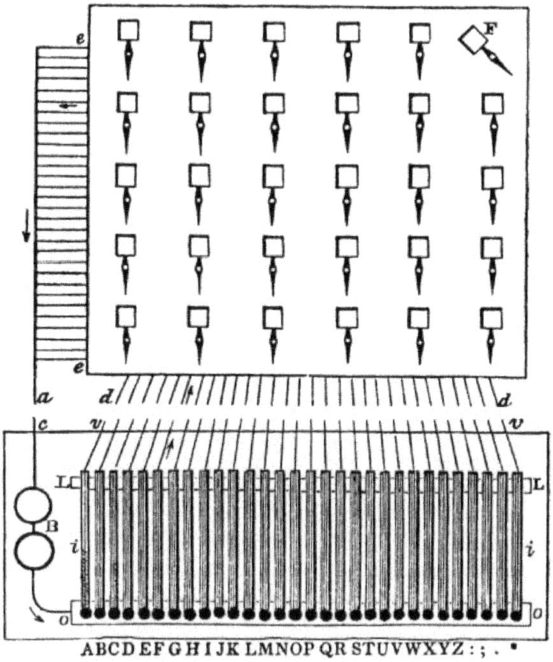

The above description is all that we have been able to obtain in relation to this plan of an electric telegraph and here introduce, figure 65, to illustrate it. The 30 needles are represented on the screen, each carrying a shade, which conceals the letter when the needle is vertical. The needle belonging to the letter F, is, however, deflected, and the letter is exposed. The screen is supposed to be at the *receiving* station. To the left hand of the screen, 30 wires, *e*, *e*, are seen joined to one, *a*; the other 30 wires, *d*, *d*, are seen below the screen. These wires may be supposed to extend many miles, and to be joined with their corresponding wires, *c*, and also *v*, *v*, of the *transmitting* station, where it will be observed, the wire, *c*, connects with the battery at one pole, and from the other pole a wire is continued and soldered to the metallic plate, *o*, *o*, which extend under all the 30 keys, *i*, *i*. These keys are each insulated, at their extremity, by being fastened to a wooden standard, L, L, to which a wire is soldered. Now, suppose the key, F, is pressed down, (the sixth key from the left,) the fluid then passes from the battery, B, through the wire to *o*, the plate; then to the key in contact with it; then to its wire, marked by the arrow; thence through the extended wire to its corresponding wire at the receiving station, denoted by the arrow; then through the coils of the multiplier, deflecting the needle, F; then returns through its wire, at the left, to the common wire, *a*; then through the extended wire to C, and the battery, of the transmitting station. In this manner any letter upon the screen may be indicated.

Extract from the Report of the Academy of Industry, in reference to a suggestion of M. Amyot of an Electric Telegraph.

"M. Amyot announced, in a letter addressed to the Academy of Sciences, in April, 1838, that he also proposed to construct an electric telegraph. It was to consist of a single current, which would move a single needle, which needle would of itself write on paper, with mathematical precision, the correspondence which might be transmitted to the other extremity, by a simple wheel on which it should be written by means of points, differently spaced, the same as they are on the barrels of portable organs. In order to send any news then, he required to write, by means of movable characters, which must be constructed in a certain manner, and immediately it would be repeated and transcribed at the place where he wished to address it, on paper, which could be put into the hands of persons specially employed to transmit despatches. But all that method of execution, which it seems ought to move is clock work, not having been sufficiently described by the author, the *most vague uncertainty* yet reigns as to the true construction of that apparatus, which appears to us to have been for M. Amyot, rather the occasion, than the end, of this communication; for indeed he attempted to make the possibility admitted of establishing a universal telegraphic language of his invention."

Edward Davy's Electric Telegraph.[34]

The following description of Mr. Davy's telegraph is taken from his specification and drawings, published in the Repertory of Patent Inventions. Although the specification has given the basis of his plan, yet the description contained therein, and the drawings representing his plan, are so obscure and deficient, that to have given it to the public in that form, would have represented it as perfectly impracticable. He has failed to state the number of signals which it is capable of giving. He has committed great errors in the arrangement of his wires for producing signals. He has introduced two keys, which produce the same signals as two others in the same arrangement. He has employed three extended wires for communicating from one station to another station, and by his arrangement of them, could not have obtained more than four signals. He has also very obscurely described his escapement, by which his marking cylinder is made to advance one division at a time for receiving the signals. This latter difficulty, however, we have been enabled to clear up, by a description of it in a work published by Mr. Bain.

Notwithstanding the imperfections and obscurities of his specification and drawings, we have endeavoured to carry out his plan, and give it a practical shape, perhaps, as Mr. Davy originally designed it.

As it is now described, there are 26 signals, or marks, indicating letters. The employment of four wires instead of three, or if Mr. Davy chooses to use for the common communicating wire the ground, which is perfectly practicable, it will reduce the number to three, the number he has specified. We have introduced one key more, and so arranged the two superfluous keys as to make them available. With this preliminary, we will proceed with the description.

Fig. 66.

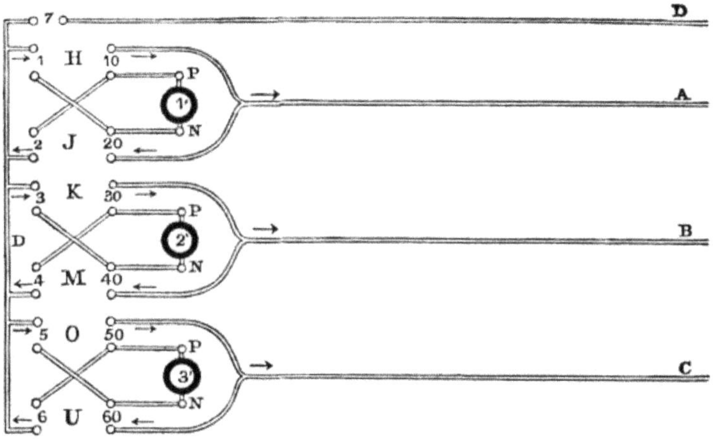

Fig. 67.

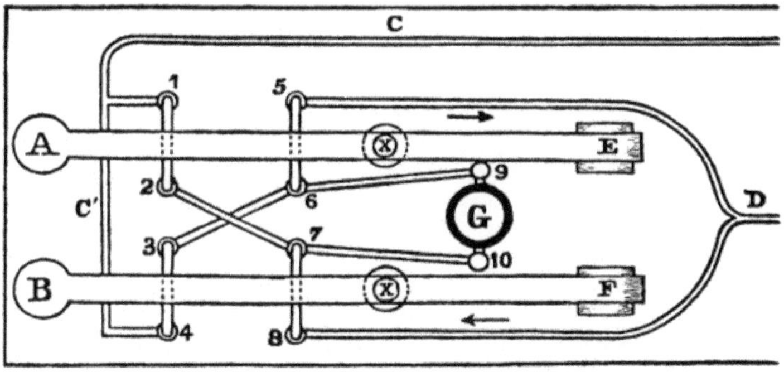

Figure 66 represents a top view of the arrangement of the wires, mercury cups, and batteries of the *transmitting station.* The close parallel lines represent the wires, of which D, A, B and C are those which proceed to the receiving station. 1', 2' and 3' are the three batteries, of which, P and N are their respective poles. The small circles formed at the termination of the wires, and marked 7, 1, 10, 2, 20, &c. are mercury cups, in which the terminating wires are immersed. The wires 1 and 20, and 2 and 10, &c. which cross each other, are not in contact, but perfectly insulated. The wires shown in this figure, are all secured permanently, with their mercury cups, to one common base board. The letters H, J, K, M, O and U represent the places of the six finger keys, used in transmitting signals. There is, also, another key at 7, for uniting the wire, D and D. In this figure, however, the keys themselves are omitted, in order to render more clear the arrangement of wires under and around them. Another figure, 67, is here introduced to illustrate the plan of one set of wires and their two keys. In figure 67 is represented, in a top view, the two wooden keys, A and B, and their axes, at E and F. G is the battery, of which, 9 is the positive pole, and 10 the negative pole. The small circles, marked 1, 2, 3, 4, 5, 6, 7 and 8 represent the mercury cups. C and C', and also, D, are the extended wires. The keys, A and B, have each two wires, passing at right angles through the wooden lever. The wires of the key, A, are marked 1 and 2, and 5 and 6, and those of the key, B, are marked 3 and 4, and 7 and 8. These wires, directly over the mercury cups, are bent down a convenient length, so as to become immersed in the cups, when the lever is depressed, and rise out of them, when the lever is elevated. Now, if the key, A, is depressed, the cup, 1, is brought in connection with cup 2; and 5 is connected with 6, by the wires, supported by the lever, being immersed in the mercury; and the key, B, not being depressed, there is no connection of the cup 3 with 4; or 7 with 8. At X and X, under the lever, are springs, which keep the lever elevated; and, consequently, the wires out of the cups, when the keys are not pressed down.

Fig. 68.

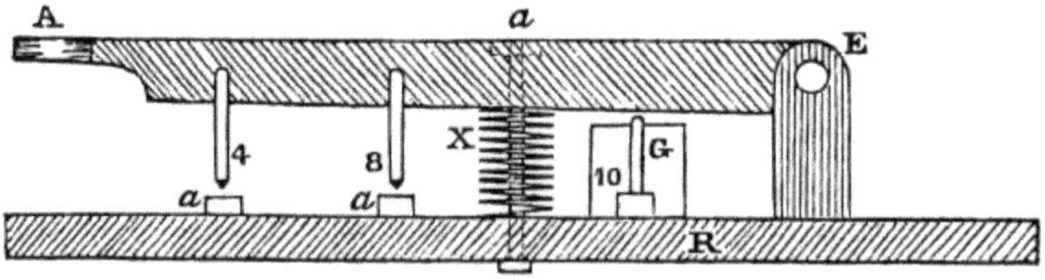

Figure 68 represents a side view of the lever, or key, A, and its axis at E. R is the platform supporting the standard of the axis; the stationary wires; the battery, G; and the mercury cups, *a, a* and 10. X is the spiral spring, for the purpose of carrying back the lever, after the finger is taken off and sustaining it in its elevated position. Through the centre of the spiral, passes a rod, with a head upon it at the top of the lever, to limit its upward motion. At its lower end, the rod is secured in the platform, R. 4 and 8 are the two wires supported by the lever, A, and are seen to project down directly over the mercury cups, *a* and *a*, so that by depressing the key, they both enter the cups and form a metallic connection. The key, B, figure 67, has the same fixtures and is similarly arranged as the key, A, represented above.

Fig. 69.

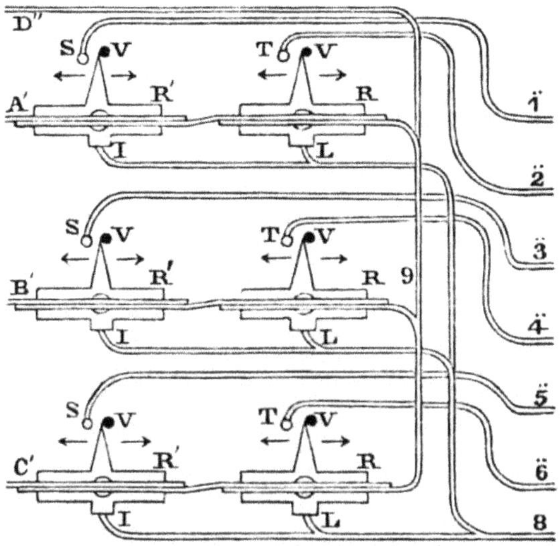

Figure 69 represents a top view of die arrangement of multipliers at the *receiving* station. R', R' and R'; R, R, and R are six magnetic needles, or bars, each of which move freely upon a vertical axis passing through their centres. The lower point of their axes is immersed in cups of mercury, in which also terminate the wires, I, I, I and L, L, L. The wires, D'', A', B' and C', are those coming from the *transmitting* station. A', B' and C', each enter the needle arrangement, and first passing from left to right, over the magnetic bars, R', R' and R', in the direction of their length, then down and under and round, making many turns, leave these three needles and pass *under* the needles, R, R and R, and in like manner from right to left round them, making a number of turns, then pass off and unite together, in the wire, 9, which is a continuation of D''. This wire is called the *common communicating wire*,[35] and the wires, A', B' and C' are called *signal wires*. At right angles, there projects from each magnetic bar, a metallic tapered arm, which rests against the studs, V, V, V, V, V, V, when the needle is undisturbed. But when the needles are made to move in the direction, to carry the arms to the left, they are brought in contact with the metallic stops, S, S, S and T, T, T. To each of these stops, it will be observed, a wire is soldered, and continued respectively from S, S, S to ¨1, ¨3, ¨5, and from T, T, T to ¨2, ¨4, ¨6. It will also be observed, that from each of the mercury cups below the magnetic bars, the wires, I and L, and I and L, and I and L, proceed and unite in pairs at, L, L, L; these three united wires are then continued, and the whole are joined in one at 8. The wires,"1, "2, "3, "4, "5, "6, are continued, in a manner hereafter to be described, and are connected with one pole of a battery. The wire, 8, is also continued and connected with the other pole. So that if any one of the needles should be made to move its arm to the left, thereby coming in contact with its metallic stop, the circuit would be complete and the current would pass along the wire,"1, for example, to the metallic stop, then to the arm, and to the magnetic bar; then to the axis; then to the mercury; then to the wire, I, and thence to the wire, 8. In the same manner the current would pass if any other arm was brought against *its* metallic stop. All the wires represented in this figure are permanently secured in their places upon a common platform.

In order to understand the combined operation of the keys and needles, figure 70 is here introduced. The right hand figure, is the same as figure 69, and the left hand the same as figure 66.

Fig. 70.
TransmittingPart of Receiving

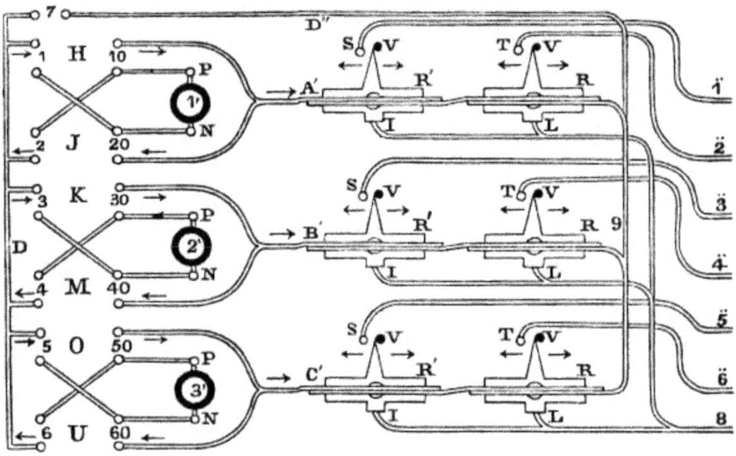

Station.Station.

The wires, D", A', B' and C', are detached from their corresponding wires of the transmitting station, and it may be imagined that many miles of wire intervene and connect the two. In the left hand figure, those mercury cups above and below, 1 and 10, are joined by two wires passing through a moving lever, in the same manner as has been described in figure 67. We will, therefore, call the key, carrying these two connecting wires, H. In like manner the key for the cups above and below the numbers, 2 and 20, is called J; for 3 and 30, is K; for 4 and 40, is M; for 5 and 50 is O; for 6 and 60, is U. The key which connects the two mercury cups on the right and left of number 7, of the wire, D'', is called 7. There are 7 keys; two for each battery, 1', 2' and 3', and each wire, A', B' and C'; and one for the common wire, D''.

It will now appear, that if the key, U and 7, are depressed, the cups above and below, numbers 6 and 60; and the cups on each side of number 7, will be connected together so that the current leaving, P, or the positive pole of the battery, 3', goes to the lower cup, 50; then by the stationary cross wire to upper cup, 6; then passes to lower cup, 6, by the wire supported by the lever, U, which is now pressed down, and its ends immersed in the two cups; then along the wire, D, to the left hand cup, 7; then to the right hand cup, 7, by the wire supported by the lever, 7, and which is immersed in the two cups; then through the extended wire to D'', of the *receiving* station; then through 9, to the two multiplying coils of the wire, C', deflecting the arm of the needle, R, to the right, against the stop, V; and the arm of the needle, R', to the left against the metallic stop, S, as indicated by the arrow at S; then along the extended wire, back to the lower cup, 60, of the *transmitting* station; then to upper cup, 60, through the wire supported by the lever, U; then to N, the negative pole of the battery, 3'.

It will be observed of the two needles, R and R', in the circuit of the same wire, C', that if R is deflected to the right against the stop, V, then R' will be deflected to the left against the metallic stop, S. The current, to produce these deflections, being through the wire C', in the contrary direction to that indicated by the arrow of the wire, C'. But if R is deflected to the left against the metallic stop, T, then R' will be deflected to the right against the stop, V. The current to produce these deflections, will then be through the wire, C', in the direction of the arrow of that wire. The same effect is produced upon the two other pairs of needles of the wires, A' and also B'. These contrary movements of the two needles, when a *current* is passing, are produced by the coils being so wound, (see figure 69,) that the wire passes round one needle in a contrary direction to what it does round the other.

If, now, we depress the keys, O and 7, the cups above and below, 5 and 50, and on each side of number 7, will be connected. The fluid will then pass from P or positive pole of the battery, 3', to the lower cup, 50; then through the key wire to upper cup, 50; then along the extended wire, C' to the *receiving* station; then through the coils of the multipliers, deflecting the arm of the needle, R, to the left against the metallic stop, T; and the arm of the needle, R', to the right against the stop, V, as indicated by the arrow at V; then to wire, 9 and D''; then along the extended wire back to the *transmitting* station, to the right hand cup, 7; then by the key wire to the left hand cup, 7; then to wire, D; then to upper cup, 5; and through the key wire to lower cup, 5; then by the cross wire to upper cup, 60, and then to N, or negative pole of the battery.

We have now shown the route of the current, when the keys, U and 7; and the keys, O and 7, were depressed. It will be observed, that when the keys, U and 7 were used, the current through the wire, D", was from *left* to *right*; and when the keys, O and 7, were used, the current was from *right* to *left*. Thus, by means of the six keys, the current of each battery may be made to pass in either direction through the *common communicating* wire, D". By the keys, U, M, J, with 7, the current is made to pass from *left* to *right* along the wire, D". By the keys, O, K, H, with 7, the current is made to pass from *right* to *left* along the wire, D". By these six keys, all those various deflections of the six needles are produced, which are necessary to close the circuit of such of the wires, ¨1, ¨2, ¨3, ¨4, ¨5, ¨6, with the wire, 8, as are required for marking the signals desired, on an instrument now to be described.

Fig. 71.

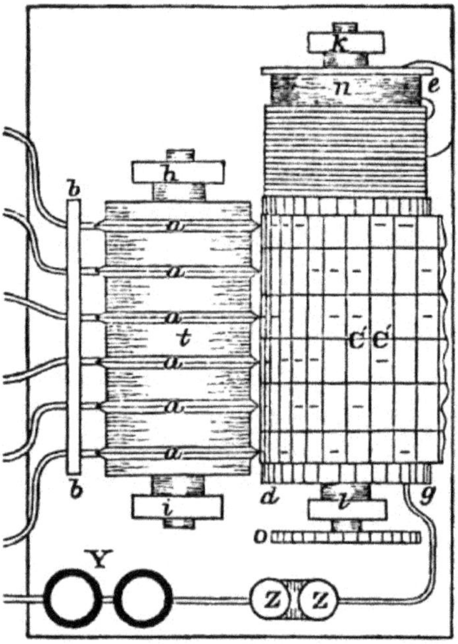

Fig. 72.

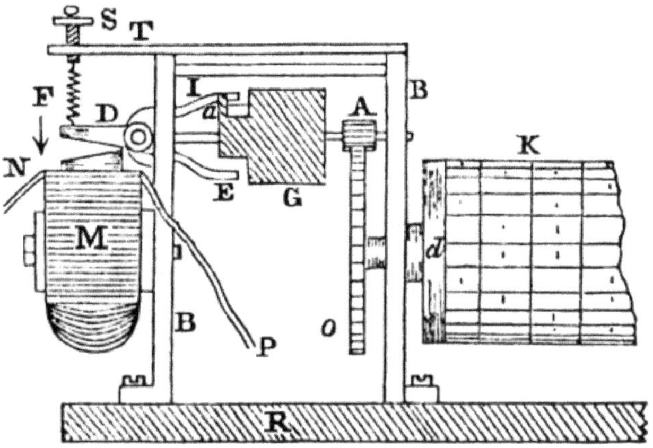

Figure 71 represents a top view of that part of the instrument at the *receiving* station, by which the signals are recorded. The seven wires on the left of the figure are a continuation of those wires, marked˙1,˙2,˙3,˙4,˙5,˙6, and 8, in figure 70. The first six pass through a wooden support, *b* and *b*, and terminate upon the edge of the platinum rings, *a*, *a*, *a*, *a*, *a* and *a*, forming a metallic contact. The six platinum rings surround a wooden insulating cylinder, *t*, which revolves upon axes in the standards, *h* and *i*. The rings are *broad* where they come in contact with the wooden roller, and are bevelled to an *edge* where they come in contact with the six wires. Y represents a compound battery, with one pole of which, wire 8, from the needle arrangement, figure 70, is connected, and from the other pole the wire proceeds to the electro magnet, Z, Z; it then passes on and is brought in connection with the metallic cylinder, *d*, at the point, *g*. The cylinder, *d*, revolves upon axes, and is supported in the standards, *k* and *l*. To the cylinder is attached a barrel, *n*, upon which is wound a cord, supporting the weight, *e*, by which the cylinder is made to revolve. C', C', represents a prepared fabric, such as calico, (impregnated with hydriodate of potass and muriate of lime,) and is placed between the platinum rings, *a*, *a*, *a*, *a*, *a*, *a*, and the metallic cylinder, *d*; *o* is a cog wheel upon the end of the axis of the cylinder, *d*, and is connected with other machinery, omitted here, but shown in figure 72, which is a side elevation of part of figure 71: *o* is the cog wheel, (figure 72,) on the arbor of the cylinder, *d*. B and B, are the two sides of the frame containing the clock work, and is secured to the platform, R: *d* is a part only of the metallic cylinder, upon which is seen a portion of the prepared fabric, K. The cog wheel, *o*, drives the pinion, A, on the shaft of the fly vane, G. M is an end view of the electro magnet, (represented by Z, Z, in figure 71,) of which N and P are the two ends of the wire composing the helix. D is its armature, constructed so as to move upon an axis represented by two small circles. To the armature are connected, and capable of moving with it, two arms, E and I, which project, so as to come in contact with the pallet, *a*, of the fly, G. F is a spiral spring, one end of which is fastened to the armature, D, and the other passes through a vertical hole in the screw, S, in the bar, T, by which the armature is held up in the position now seen, when not attracted by the electro magnet. Now, if the wires, N and P, connected with battery, Y, (figure 71,) have their circuit closed, the current passing through the helix of the magnet, M, brings down the armature, D, in the direction of the arrow, which raises the arm, I, against which the pallet, *a*, of the fly vane, is resting, and releases the fly. It then makes a half revolution and is again arrested by the pallet against the lower arm, E, and the cylinder, *d*, with its fabric, has advanced a half division. If the circuit is now broken, the armature, D, is carried up by the spring, F, at the same time the arm, E, releases the pallet, *a*, and the fly makes another half revolution, and is again stopped by the arm, I. The cylinder has now made another advance of half a division, which, together, makes a whole division the fabric has advanced. The purposes for which this is designed will now be described.

Fig. 73.

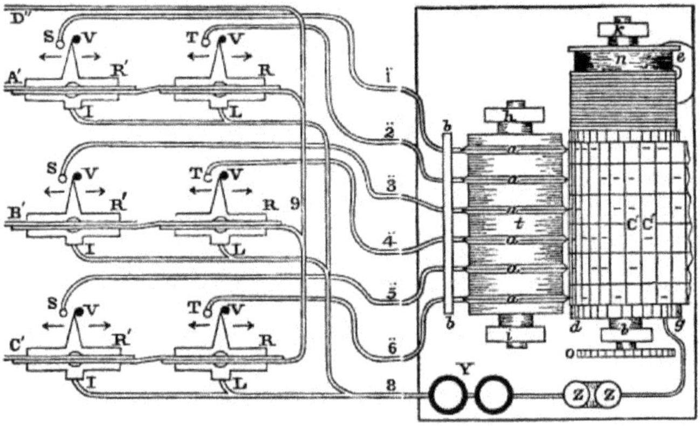

Figure 73 represents a top view of the whole apparatus of the *receiving* station. The fabric, C', C', is marked in equal divisions across it, and in six equal divisions, in the directions of its length, thus marking it into squares. Each platinum ring, *a*, *a*, *a*, &c. (when the instrument is not in operation,) is in contact with the fabric at the *middle* of the squares across the fabric. It will be observed, that the wires '1, '2, '3, '4, '5, '6 are in connection with the battery, Y, and the circuit complete, except at the arms of the needles. Suppose, for example, the arm of the needle, R', of the wire, C', is brought up against the stop of the wire, '5, at S; the circuit is then closed, and the current leaves the battery, and passes to the electro magnet, (causing the cylinder and fabric to move half a division,) then to the metallic cylinder, *d*; then through the fabric, *c'*, *c'*, resting upon the cylinder, (where it is in contact with the platinum ring, *a*, of the wire, 5,) then to the platinum ring; then to wire 5; then to the metallic stop, S; then to the arm of the needle, R', along its axis to the mercury; then to the wire, I; then to wire, 8, and to the other pole of the battery, Y. Thus a current is passed through the prepared fabric, and a mark produced thereon, in the middle of its square. If the circuit is now broken, the cylinder moves another half division, which will bring the rings to the centre of the squares, ready for the next signal.

But one battery, Y, is used for all the six circuits, formed with the wire, 8; so that, when three of the circuits are closed at the same instant, as will be shown hereafter, the current passes through the three wires of their respective circuits, making each their appropriate mark upon the fabric.

We now proceed to describe the manner of operating with the two instruments, at their respective stations: and, first, we must here designate each needle by its own peculiar mark of reference. Let the two needles upon the wire, A', be denoted by, A, S and A, T; those of the wire, B', by B, S and B, T; and those of the wire, C', by C, S and C, T. It will appear obvious, from the foregoing description, that but *one* needle of each *wire*, A', B', C', can be made to close its circuit at the same instant. However, *two* needles, or *three* needles of *different wires*, may close their circuits at the same instant, but no higher number than three. The various combinations of *one* mark, *two* marks, and *three* marks, upon the same row of six cross divisions of the fabric, constitute the characters representing letters.

Fig. 74.

LONDON.—*Transmitting Station.*

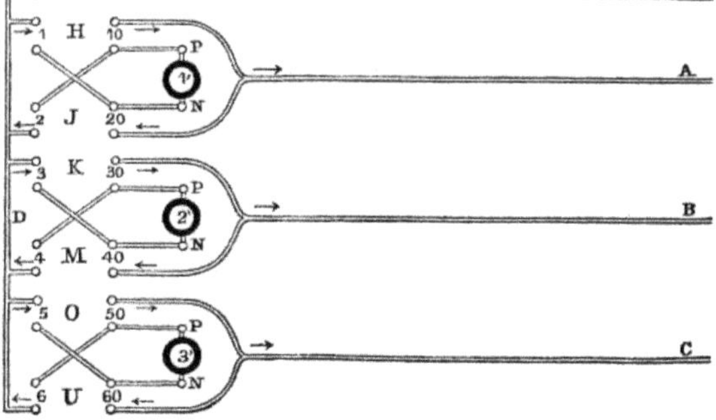

Figure 74 represents the *transmitting* station, which may be supposed to be *London*, and figure 75, the *receiving* station, which may be at *Birmingham*, with four wires extending from station to station, or three only, if the *ground* be substituted for the wire, D, D''. The wires, D, A, B and C, are supposed to be united with D'', A', B' and C', respectively. Now, if we depress the keys, in the following order, we shall, for each key, have the following deflections of the two needles, belonging to each key.

No.1.

The keys,	H, 7,	moves the arm,	A, S,	to the	right,	A, T,	to the	left.
"	J, 7,	"	A, S,	"	left,	A, T,	"	right.
"	K, 7,	"	B, S,	"	right,	B, T,	"	left.
"	M, 7,	"	B, S,	"	left,	B, T,	"	right.
"	O, 7,	"	C, S,	"	right,	C, T,	"	left.
"	U, 7,	"	C, S,	"	left,	C, T,	"	right.

These are all the various deflections which it is possible to give the six needles. Those, however, which deflect to the right, not closing the circuit, produce no effect, and are of no account. We will, therefore, omit them, and simply give the table, thus:

No.2.

The keys,	H, 7,	move the arm	A, T,	to the left.	No.	1.
"	J, 7,	"	A, S,	"	"	2.
"	K, 7,	"	B, T,	"	"	3.
"	M, 7,	"	B, S,	"	"	4.
"	O, 7,	"	C, T,	"	"	5.
"	U, 7,	"	C, S,	"	"	6.

Fig. 75.

BIRMINGHAM.—*Receiving Station.*

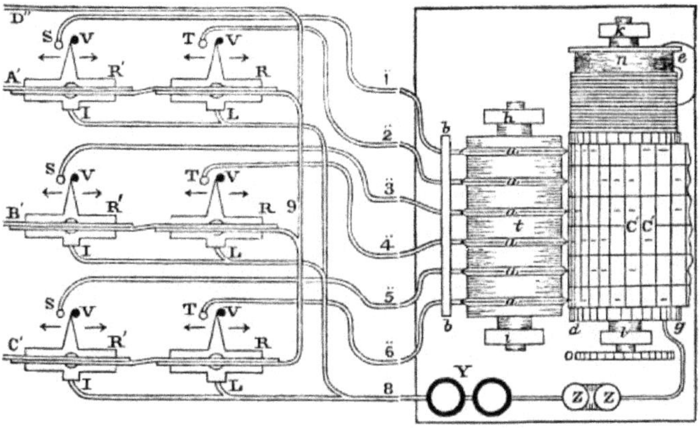

In the following table, the first column represents the keys, which when depressed, produce a deflection of the needles, (represented in the columns, second, third and fourth,) by means of their batteries, and thus closing the circuit of the wires, ˙1, ˙2, ˙3, ¨4, ˙5 and ¨6, by which the fluid, is made to pass through the prepared fabric, and mark upon its space, or spaces, numbered 1, 2, 3, 4, 5 and 6, in the fifth column. In the sixth column are the letters which the marks upon the fabric are intended to represent.

Keys.	**Needles.**	**Needles.**	**Needles.**	**Spaces on Fabric.**	**Letters.**
H, 7,	A, T,	-	-	1,	A.
J, 7,	A, S,	-	-	2,	B.
K, 7,	B, T,	-	-	3,	C.
M, 7,	B, S,	-	-	4,	D.
O, 7,	C, T,	-	-	5,	E.
U, 7,	C, S,	-	-	6,	F.
H, K, 7,	A, T,	B, T,	-	1, 3,	G.
J, M, 7,	A, S,	B, S,	-	2, 4,	H.
K, O, 7,	B, T,	C, T,	-	3, 5,	I.
M, U, 7,	B, S,	C, S,	-	4, 6,	J.

Keys.	Needles.	Needles.	Needles.	Spaces on Fabric.	Letters.
H, O, 7,	A, T,	C, T,	-	1, 5,	K.
J, U, 7,	A, S,	C, S,	-	2, 6,	L.
H, M,	A, T,	B, S,	-	1, 4,	M.
J, K,	A, S,	B, T,	-	2, 3,	N.
K, U,	B, T,	C, S,	-	3, 6,	O.
M, O,	B, S,	C, T,	-	4, 5,	P.
H,U,	A, T,	C, S,	-	1, 6,	Q.
J, O,	A, S,	C, T,	-	2, 5,	R.
H, K, O, 7,	A, T,	B, T,	C, T,	1, 3, 5,	S.
J, M, U, 7,	A, S,	B, S,	C, S,	2, 4, 6,	T.
H, K, U,	A, T,	B, T,	C, S,	1, 3, 6,	U.
J, M, O,	A, S,	B, S,	C, T,	2, 4, 5,	V.
H, M, U,	A, T,	B, S,	C, S,	1, 4, 6,	W.
J, K, U,	A, S,	B, T,	C, S,	2, 3, 6,	X.
H, M, O,	A, T,	B, S,	C, T,	1, 4, 5,	Y.
J, K, O,	A, S,	B, T,	C, T,	2, 3, 5,	Z.

Telegraphic Letters.

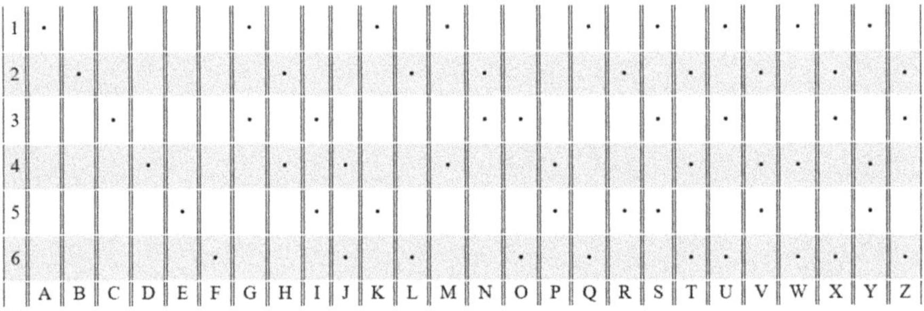

The above represents the telegraphic characters marked upon the prepared fabric. The spaces are numbered from the top.

The first six of the telegraphic letters require each a signal wire, and the common wire, D, with one battery.

The next six require each two signal wires, with two batteries, whose joint currents pass in the same direction on the common wire, D.

The next six require each two signal wires only, with two batteries, joined together so as to form a compound battery. The negative pole of one, connected with the positive pole of the other.

The next two require each three signal wires, with three batteries, whose joint currents pass in the same direction along the common wire, D.

The next six require each, three *signal* wires only, with three batteries. One of the signal wires with its battery is used as a common wire for the other two. Hence the current of the two batteries of the two signal wires unite in one, and are connected with the battery of the common wire as a compound battery.

With what rapidity these letters may be formed, does not appear, or to what extent the plan has been carried out.

Bain's Printing Telegraph.

The following description of Mr. Bain's plan of what he calls an *electro magnetic* printing telegraph, is taken from a work entitled, "An account of some remarkable applications of the electric fluid to the useful arts, by Alexander Bain. Edited by John Finlaison, Esq. London, 1843."

It appears from this work that Mr. Bain's plan was invented in 1840, and the following certificate is given in reference to the date of its first operation.

PERCEIVAL STREET, CLERKENWELL, *Aug. 28, 1842.*

DEAR SIR—In reference to your application, I recollect visiting you at your apartments in Wigmore street, early in July, 1840, when you showed me the model of your *electro magnetic* printing telegraph, with which you printed my name at the time. You also showed me a model of your electro magnetic clock, and explained to me the principles and utility of them.

I remain, dear sir, yours, respectfully,
ROBERT C. PINKERTON.

To MR. ALEXANDER BAIN.

Fig. 76.

PORTSMOUTH.

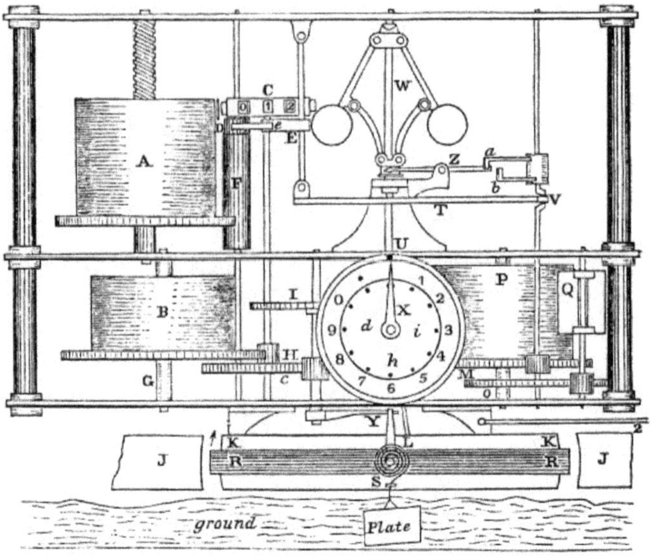

Fig. 77.

LONDON.

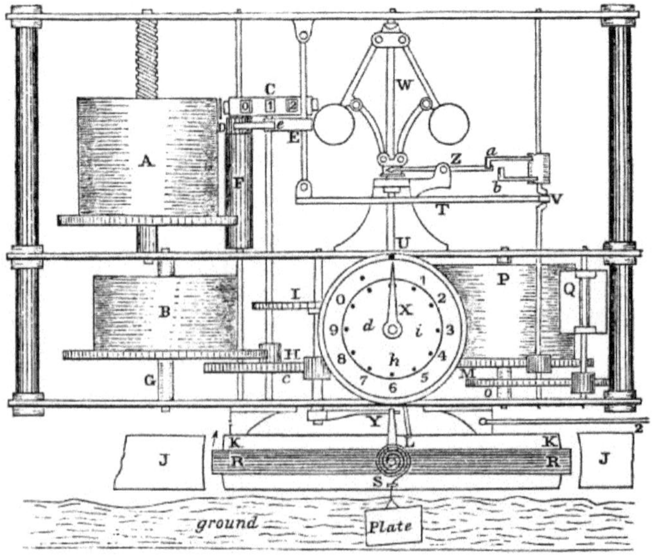

Figures 76 and 77 exhibit the arrangements of Mr. Bain's telegraph. Both figures are the same, representing one as being at *Portsmouth*, and the other at *London*. The same letters will refer to either instrument: *d*, *i* and *h*, represent the signal dials, insulated from the machine. X is a hand or pointer. The small dots represent twelve holes in the dial, corresponding with the twelve signals, and two blanks, 1, 2, 3, 4, 5, 6, 7, 8, 9, 0. U is a similar hole over the starting point of the hand, X. R is a coil of wire, freely suspended on centres. K and K, is a compound permanent magnet, placed within the coil, and immovably fixed upon the frame of the machine. J and J are sections of similar permanent magnets. S is a spiral spring, (and there is another on the opposite side,) which conveys the electric current to the wire coil, and at the same time leaves the coil free to move in obedience to the magnetic influence. So long as the electricity is passing, the wire coil continues to be deflected, but the instant the electric current is broken, the springs, S, bring back the coil to its *natural position.*[36] L is an arm fixed to, and carried by the wire coil, R and R, to stop the rotation of the machinery. B is a main spring barrel, acting on the train of wheels, G, H and I, which communicate motion to the governor, W, and the hand, X. On the arbor of the wheel, H, is fixed a type wheel, C, at a little distance from the paper cylinder, A, on which the messages are to be imprinted. P is a second main spring barrel, with its train of wheels, M, O. Q, is a fly, or vane. On the arbor of the wheel, *o*, there is a crank, V, and two pallets, *a* and *b*, which prevent the train of wheels from rotating, by coming in contact with the lever, Z. When the telegraph is not at work, a current of electricity is constantly passing from the *Portsmouth plate*, buried in the ground, through the moisture of the earth, to the plate in the ground at the *London* station. From the copper plate of that station the electric current passes up through the freely suspended multiplying coil, R and R, (which it deflects to the horizontal position,) into the machinery, and thence to the dial, by means of a metal pin, inserted in the hole, U; from the dial it passes by a single insulated conducting wire, 1, suspended in the air, back to the first machine; traversing which, it passes through the freely suspended multiplied coil, R and R, which it deflects, also, to the horizontal position to the plate from whence it started, and thus completes the circuit.

"When a communication is to be transmitted from either end of the line (one station only being able to transmit at a time,) the operator draws out the metal pin from the hole, U, in the dial of his machine; the electric circuit is then broken, and the ends of the multiplying coils, R and R, at both stations are carried upwards, in the direction of the arrow, by the force of the spiral springs. The arms, L, attached to the two coils, moving to the right, release the lever, Y, which leaves the machinery free to rotate, and as the moving and regulating powers are the same at both places,[37] the machines go accurately together; that is, the hands of both machines pass over similar signals at the same *instant* of time, and similar types are continually brought opposite to the printing cylinders at the same moment. An inspection of the wheel work will show, that this movement will have caused the governor, W, to make several revolutions, and the divergence of the balls, in obedience to centrifugal force, will have raised one end of the lever, Z, and depressed the other, which allows the pallet, *a*, to escape; but the rotation of the arbor is still opposed by contact with the second pallet, *b*. The operator having inserted the metal pin in the hole, under the signal which he wishes to communicate, the moment the hand of the dial comes in contact with it, the circuit is again completed, and both machines are stopped instantly. The governor balls, collapsing, depress the left hand end of the lever, Z, clear the pallet, *b*, and this allows the crank spindle, V, to make one revolution.

"The motion of the crank by means of the crank rod, T, acting on the lever, E, presses the type against the paper cylinder, A, and leaves an impress upon the paper; at the same time, a spring, *e*, attached to an arm of the lever, E, takes into a tooth of the small ratchet wheel, D, on the spindle of the long pinion, F, which takes into and drives the cylinder wheel; so that the crank apparatus, going back to its former position, after impressing a letter, moves the signal cylinder forward, and presents a fresh surface to the action of the next type. As the cylinder moves round, it has also a spiral motion upward, which causes the message to be printed in a continuous spiral line until the cylinder is filled.[38] In order to mark, in a distinct and legible manner, the letters printed by the apparatus, two thicknesses of riband, saturated with printing ink and dyed, are supported by two rollers so as to interpose between the type wheel and the cylinder; (the rollers are not shown in the figure, to prevent confusion.) If a second copy of the message, thus simultaneously printed at two distant places, is desired at either, a slip of white paper is placed between the ribands to receive the imprint at the same time as the cylinder."

Fig. 78.

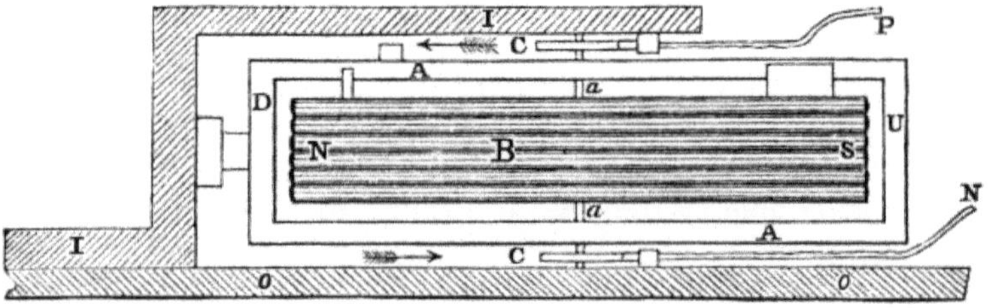

Figure 78 represents a top view of the coil and magnets of Mr. Bain's machine. B is the compound permanent magnet, with six bars. N is the north pole, and S the south pole. A, A are the sides of the brass frame containing the coils; C, C are the spiral springs on each side: *a* and *a* is the axis of the coil: *o*, *o*, is a part of the frame containing the clock work, (not shown in this figure,) supporting one centre of the coil, and I and I a support for the other centre. N and P are the wires, one of which is in connection with the ground, and the other with the extended wire. When the circuit is closed, and the current from P pole of the battery is in the direction of the arrow above, and then through the coil to the other pole, N, in the direction of the arrow below; the end, D, of the coil, will be depressed, and the end, U, will rise; reverse the current and the effect is the elevation of the end, D, of the coil, and the depression of the end, U.

Wheatstone's Rotating Disc Telegraph, invented, 1841.

Figure 79 represents that portion of the instrument which belongs to the *transmitting* station, of which, K, is a circular disc, with the alphabet and numerals, marked in two concentric circles upon it: *a* are handles projecting from its rim, one to every letter, by means of which, the disc is turned upon its axis, and brought to that position, *b*, required for signalizing a letter. O is a side view of the disc, K: *t* is the rim of the disc, with its holders: *h* is a portion of the axis of the disc, shown as broken off: *c* represents a silver band surrounding a pulley, or hub, upon the axis, and directly behind the disc. Upon the hub are metallic ribs, *b*, parallel with its axis, corresponding in number to the letters on the dial. Each rib forms a metallic contact with the silver band, *c*, and are separated from each other by pieces of ivory, fastened to the hub. Both the ribs and ivory pieces are made perfectly smooth and even upon their surface: *e* is a metallic spring with a portion of it pressing against that portion of the hub between the silver band, *c*, and the disc, *t*, in such a manner that when the disc is turned, the metallic ribs and ivory pieces shall alternately come in contact with it. To this spring is soldered a wire connected with one pole of the battery, *g*, and from the other pole proceeds the wire, *n*: *d* is another metallic spring, similar to *e*, but pressing *only* upon the silver band, with which it is always in contact, and to which a wire, *p*, is soldered. Whenever the spring, *e*, is in contact with any of the metallic ribs, there is a continuous connection from *n* to *p*, viz. from *p*, to the spring in contact with the silver band, *c*, thence to the rib with which the spring, *e*, is in contact; then to the spring, *e*, then to the battery, *g*, and then to the wire, *n*. If, however, the disc, O, should be turned, so that the spring, *e*, is in contact with the ivory, then the circuit is broken at that point, and in this manner the circuit is alternately broken and closed as the wheel, O, is turned from one letter to another by means of the handles at *t*.

Fig. 79.

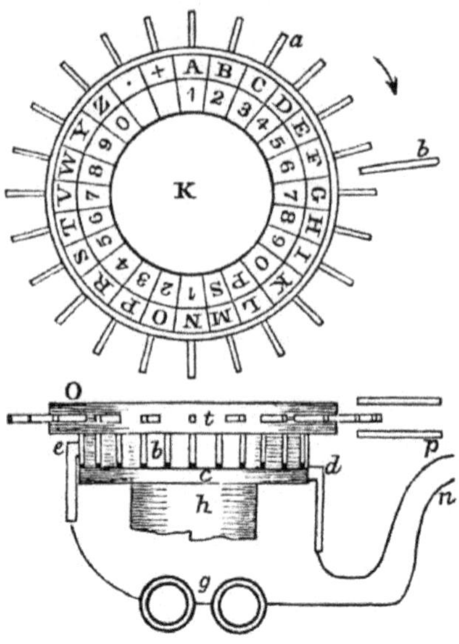

Fig. 80.

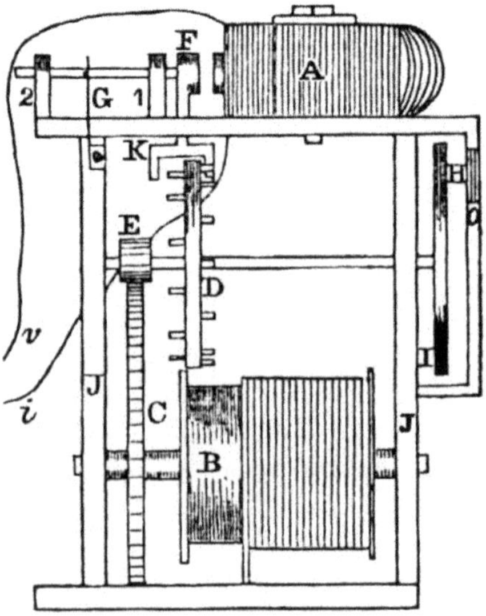

Figure 80 represents a side elevation of the dial and clock work of the *receiving station.* A represents an edge view of the electro magnet, from which proceed the two wires, *v* and *i*, which connect with the wires, *n* and *p*, of figure 79. J and J is the brass frame containing the wheel work, C and E; the pin wheel, D; the dial plate, I; and the barrel, B, which is driven by a weight and cord. In the side of the wheel, D, are pins projecting from the rim, parallel with the axis, and are equal in number to the divisions, or letters, upon the dial, I. They are, however, placed alternately on each side of the rim. F is the armature of the magnet, fastened upon a horizontal rod, sliding freely through the standards, 1 and 2. G represents a spring, fastened to the frame, J, and which carries back the armature, F, when the magnet has ceased to attract it. From the armature there extends downward an arm, K, which, as it approaches the pin wheel, D, presents two arms, or pallets, one on each side of the wheel. These pallets are so arranged with regard to the pins, that if one pallet releases a pin on one side of the wheel, the same movement will cause the other pallet on the other side, to arrest the motion of the wheel by its striking against the next alternate pin. H and I is an edge view of the circular dial, enclosed in a case, with a single opening at O, so that only one letter at a time can be seen. This dial, I, is in every respect marked as the disc in figure 79.

Figure 81 represents the two instruments. O the *transmitting* instrument, and the right hand figure the *receiving* instrument. The wires, *v* and *i*, are respectively connected with *p* and *n*. It will be observed, that the armature, F, is not attracted, and that the right hand pallet is checking the pin wheel, so that the dial is stationary. If, however, the disc, *t*, is turned so that the circuit is completed, by the contact of the spring, *e*, with one of the ribs, instantly the armature is attracted by the electro magnet, which will carry the right hand pallet away from the pin wheel, and which will then move by the action of the weight upon the barrel, B, until it is checked by the left hand pallet, which had advanced to the wheel at the same time the other receded. This single operation has moved the disc one division and the armature is still attracted. Now let the disc, *o*, be turned until the spring, *e*, has been passed by the rib, and is in contact with the ivory only, instantly the current ceases; the armature, F, recedes from the magnet by the action of the spring, G; this has taken the left hand pallet from the pin wheel, which is permitted to move until the next pin strikes against the right hand pallet. This has now brought another letter in front of the aperture at H. Thus it will be seen, that the design of this instrument is to bring into view, at the aperture such letters as are required in transmitting a message.

Fig. 81.

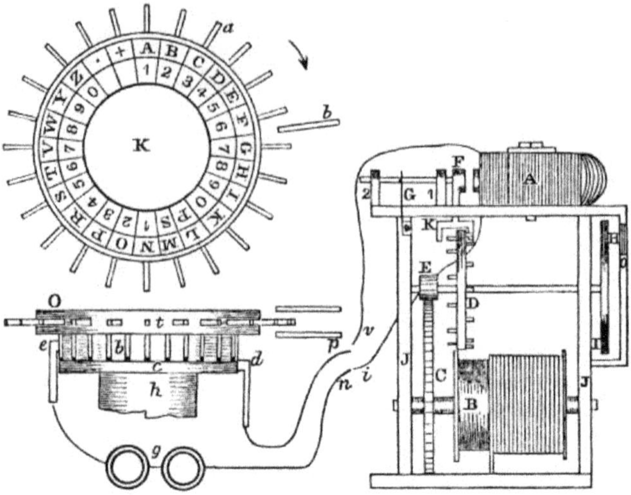

Suppose letter A, is at the point, *b*, of the *disc*; and letter A of the *dial* is opposite the opening; the instrument is now ready to transmit, and let the letter, I, be the first of the message. The operator gently turns the disc round in the direction of the arrow, so that each time the circuit is broken a new letter appears at the dial, and each time it is closed by the operation of the pallets, in checking and releasing the pin wheel. This is its operation until the letter, I, has reached the point, *b*, when a short pause is made. The next letter, H, requires but one movement of the disc, then follows, A; then, V; and then, E.

In relation to this instrument, Professor Daniell says: "We can only further briefly allude to two of the most important modifications of this invention, which Prof. Wheatstone has made for specific purposes. By substituting for the paper disc, on the circumference of which the letters are printed, a thin disc of brass, cut from the circumference to the centre, so as to form 24 springs, on the extremities of which, types, or punches, are placed, and adding a mechanism the detent of which, acted on by an electro magnet, causes a hammer to strike the punch against a cylinder, round which are rolled, alternately, several sheets of white paper, and of the blackened paper used in the manifold writing apparatus, he has been enabled to obtain, without presenting any resistance to the type wheel, several distinct printed copies at the same time of the message transmitted."[39]

Mr. Wheatstone has recently so modified his telegraph as to use two needles, or galvanometers, and two extended wires, with the ground as half the circuit for the two wires. He has thus adopted PROF. MORSE'S *plan* of using the ground as a common conductor for two or more wires. He, however, still requires two wires for *one* independent line of communication; one station only being able to communicate at a same time. He has no mode of recording his message, but depends upon the watchful eye of the attendant. His code of signals are based upon Schilling's plan, heretofore described, page 155, and also Gauss and Weber's, page 156, from whom he seems to have obtained his idea.

The two needles, or galvanometers, stand side by side, one of which is called the *left* needle and the other the *right* needle. These two needles are placed directly in front of the person who transmits. There are, also, in front, two handles, one for each hand, with which the operator transmits a message, closing and breaking the circuit of the two wires. His signals are made thus: The upper half of the left hand needle moving to the left twice, gives, *a*; three times, *b*; once to the right and once to the left, *c*; once to the left and once to the right, *d*; and, in like manner, for the other letters of the alphabet, as shown in the table which follows.

Left Hand Needle.				Right Hand Needle.			
ll,	A.	r,	E.	l,	H.	lr,	M.
lll,	B.	rr,	F.	ll,	I.	r,	N.
rl,	C.	rrr,	G.	lll,	K.	rr,	O.
lr,	D.			rl,	L.	rrr,	P.

Joint Action of Both Needles.

l,		l,	R.
ll,		ll,	S.
lll,		lll,	T.
rl,		rl,	U.
r,		r,	W.
rr,		rr,	X.
rrr,		rrr,	Y.
r,	completed.		

ll,	rr,	I understand, or yes.			
rl,	rl,	I do not understand, or no.			
rl,	rl,	1.			
lr,	lr,	2.			
r,	r,	3.			
			l,	l,	4.
			rl,	rl,	5.
			lr,	lr,	6.
			r,	r,	7.
l,	l,		l,	l,	8.
ll,	ll,		ll,	ll,	9.
r,	r,		r,	r,	0.

Mr. Wheatstone does not appear to be aware of all the advantages of this, his latest plan of using two needles and two wires, since some of his signals for the *numerals*, are repetitions of his *letter* signals, and require four deflections of a single needle, with a pause between the two first deflections, and the two last, and for *some* other signals he requires as many as three deflections of a signal needle. He has likewise, apparently, for want of simple signals, omitted the letters, J, Q, V, Z. He could with perfect ease, obtain from his two wires and two needles, sixty-four different signals, requiring the time of only two deflections, each, and using but one hand for manipulating four keys, instead of both hands, as in his present plan. The author has demonstrated it by actual experiment.

Footnotes:

[1] These are made at the American Pottery, in Jersey City, opposite New York.

[2] The term *magnet*, here, is synonymously used with the iron for the magnet, as the simple iron is not a magnet, except when subjected to the action of the battery through the helices of wire around it. It would confuse the reader, if this distinction be not kept in view. *Permanent magnets* are those which retain their magnetism when once they are charged. They are always made of steel, and usually bent in the form of a horse-shoe. Sometimes they are of a single plate of that form, and others are constructed with many plates, side by side,

fastened together so as to present a compact magnet of the same form. They are distinguished from *Electro Magnets* from the fact, that the soft iron of the latter depends upon the influence of the galvanic fluid for its magnetism, and retains it only so long as the soft iron is under its influence, while the former, when once submitted to the influence of the galvanic fluid, retain their magnetism permanently.

[3] One marking point will suffice.

[4] The paper used for telegraphic writing is first manufactured by the paper making machine in one long continuous sheet, of any length, about three feet and a half in width, and is compactly rolled up as it is made, upon a wooden cylinder. It is then put into a lathe and marked off in equal divisions of one and a half inches in width; a knife is applied to one division at a time, and as the roll of paper revolves, the knife cuts through the entire coil until it reaches the wooden centre. This furnishes a coil ready for the register, and is about fifteen inches in diameter. The whole roll of paper furnishes, in this way, about twenty-eight small rolls prepared for use.

[5] The pulley and cord have been dispensed with and two small cog wheels substituted.

[6] At this time the key is opened at the station from which the communication is to be sent.

[7] The first working model of the Telegraph was furnished with a lead pencil, for writing its characters upon paper. This was found to require too much attention, as it needed frequent sharpening, and in other respects was found inferior to a pen of peculiar construction, which was afterwards substituted. This pen was supplied with ink from a reservoir attached to it. It answered well, so long as care was taken to keep up a proper supply of ink, which, from the character of the letters, and sometimes the rapid, and at others the slow rate of writing, was found to be difficult and troublesome. And then again, if the pen ceased writing for a little time, the ink evaporated and left a sediment in the pen, requiring it to be cleaned, before it was again in writing order. These difficulties turned the attention of the inventor to other modes of writing, differing from the two previous modes. A variety of experiments were made, and among them, one upon the principle of the manifold letter writers; and which answered the purpose very well, for a short time. This plan was also found objectionable, and after much time and expense expended upon it, it was thrown aside for the present mode of marking the telegraphic letter. This mode has been found to answer in every respect all that could be desired. It produces an impression upon the paper, not to be mistaken. It is clean, and the points making the impression being of the very hardest steel, do not wear, and renders the writing apparatus always ready for use.

[8] See Silliman's Journal, vol. 35, 1839, pages 258-267.

[9] Franklin appears to have been the first, or among the first, who used the ground as part of a conducting circuit in the performance of electrical experiments. Steinheil it appears was the first to use the ground as a conductor for magneto electricity. Bain, in 1840, was the first to use the ground as a source of electricity in conjunction with its conducting power, as a circuit. Prof. Morse, has since the establishment of the telegraphic line, used the ground as half the line, with perfect success, employing the battery; and Mr. Vail, in an experiment in 1844, succeeded in operating the electro magnet, with its armature attached to a lever, without any battery.

[10] In Prof. Daniel's, Introduction to the Study of Chemical Philosophy, 2d edition, 1843, there are these facts to be noticed. In the preface, there are these words: "It only remains for me now, to acknowledge my obligations to my friends and colleagues, *Professor Wheatstone* and Dr. Todd, for their great kindness in undergoing the disagreeable labour of revising and correcting the proof sheets. They have thereby prevented many errors which would have otherwise deformed the work."

No statement then of Prof. Daniel's, particularly in that part of his work which related especially to Wheatstone's Telegraph, would be allowed to pass unnoticed by Mr. Wheatstone and we are authorizsed in considering any such statement as having his sanction.

We then find, page 576, the following statement: "Ingenious as Prof. Wheatstone's, contrivances are, they would have been of no avail for telegraphic purposes, without the investigation which he was the first to make of the laws of electro magnets, when acted on through great lengths of wire. *Electro magnets of the greatest power, even when the most energetic batteries are employed, utterly cease to act when they are connected by considerable lengths of wire with the battery.*"

If any thing were needed to show that Prof. Wheatstone was not the inventor of the *Electro Magnetic Telegraph*, it is this assertion (under the supervision of Prof. Wheatstone) made by Prof. Daniel. In 1843, Prof. Wheatstone had not made the discovery upon which Prof. Morse bases his invention, viz. that *Electro Magnets can be made to act, with an inconsiderable battery too, when the latter is connected with the former by considerable lengths of wire*: 80 miles may certainly be considered as of *considerable length.*

[11] It now occupies a space 10 inches long, 8 inches high, and 5 wide.

[12] Mr. Francis O. J. Smith has recently published a Secret Corresponding Vocabulary adapted to this purpose.

[13] It is proper that I should here state, that the patent-right is now jointly owned, in unequal shares, by myself, Prof. Gale of New York City University, and Messrs. Alfred and George Vail.

[14] This line could now be constructed for less than half the sum.

[15] 98, per minute, can now be sent, 1845.

[16] Many of the facts here given, are taken from Priestley's Work upon Electricity.

[17] "As the possibility of this experiment has not been easily conceived, I shall here describe it. Two iron rods, about three feet long, were planted just within the margin of the river, on the opposite sides. A thick piece of wire, with a small round knob at its end, was fixed on the top of one of the rods, bending downwards, so as to deliver commodiously the spark upon the surface of the spirit. A small wire, fastened by one end to the handle of the spoon containing the spirit, was carried across the river, and supported in the air by the rope commonly used to hold by, in drawing ferry boats over. The other end of this wire was tied round the coating of the bottle; which being charged, the spark was delivered from the hook to the top of the rod standing in the water on that side. At the same instant the rod on the other side delivered a spark into the spoon and fired the spirit; the electric fire returning to the coating of the bottle,through the handle of the spoon and the supported wire connected with them."

[18] "An electrified bumper is a small thin glass tumbler, nearly filled with wine, and electrified as the bottle. This, when brought to the lips, gives a shock, if the party be close shaved, and does not breathe on the liquor."

[19] Academy of Sciences at Munich.

[20] Encyclopedia Britannica, vol. 21, p. 686.

[21] Report of Academy of Industry, Paris.

[22] Polytechnic Central Journal, 1838.

[23] We here introduce to the reader our ingenious and scientific country man, Mr. Joseph Saxton, formerly of the United States mint, Philadelphia, but now connected with the Department of weights and measures, at Washington, who invented the first Rotary Magneto Electric Machine, and which has now been extensively adopted.

[24] M. M. Nobili and Antinori.

[25] Mr. Saxton on the 3d of May exhibited his apparatus, and the mode of obtaining the spark to Dr. Ritchie, Messrs. Thomas Gill, John Isaac Hawkens and Steadman Whitwell. On the 8th of May he loaned it to Dr. Ritchie, who

publicly exhibited it at a lecture, at the London University, and also at the London Institution, Finsbury.

[26] In relation to this instrument, Prof. Daniell makes the following remarks: "After Dr. Faraday's discovery of *Volta electric* and *magneto electric* induction, many ingenious contrivances were made for exalting the effects and facilitating experiments. The most complete arrangement now in use, was the original combination of Mr. Saxton."

[27] From the Polytechnic Central Journal, 1838, Nos. 31, 32.

[28] From the Polytechnic Central Journal, 1838.

[29] A day's work of a fair compositor in setting up type is 6,000 ems, equivalent to 12,000 pieces, in ten hours, or 20 pieces per minute. A very quick and expert compositor may set up 10,000 in the same time, equal to 20,000 pieces, or 33⅓ pieces per minute. One em is equivalent to about two pieces.

[30] The author has recently devised a new plan for printing with type, in which the pendulum movement is dispensed with, and the motion of the type wheel is dependent upon the control and government of certain apparatus at the transmitting station. This controlling part is capable of giving to the type wheel a most rapid movement, and from an estimate made from some actual tests, the number of letters capable of being printed, are increased much beyond the former plan, taking the message already used as an example. Still he considers it inferior to that mode, now adopted by Professor Morse.

[31] Mr. Vail invented an instrument with this arrangement 16 years ago, for the purpose of printing speeches as fast as delivered.

[32] Steinheil in the account he gives of his own telegraph, says, "Gauss mentions a communication from Humboldt, according to which Belancourt, in 1798, established a communication between Madrid and Aranjuez, a distance of 26 miles, by means of a wire, through which a Leyden jar used to be discharged, which was intended to be used as a telegraphic signal."

[33] Report of the Academy of Industry, Paris, 1839.

[34] From the Repertory of Patent Inventions, No. lxvii. New Series, London, July, 1839.—Sealed, July 4th, 1888.

[35] A', B' and C' are also, occasionally, common communicating wires.

[36] Mr. Bain means, by the *deflected position* of the coil, (when the current is passing,) its *horizontal* position, as shown in the figure. Its *natural* position, (when the current is broken,) is the elevation of the left hand end of the coil, in the direction of the arrow, carried up by the power of the spring, at the

centre of the coil. This action of the spring is overcome, when the current is passing, to such a degree, as to bring the coil to the horizontal position as represented in the figure.

[37] It is absolutely necessary to the certain and accurate performance of the two machines, that their movements should be synchronical, or else a different figure, or signal, from that intended by the operator at the transmitting station, may be given at the receiving station.

[38] This contrivance for moving the paper is exactly similar to that in Prof. Morse's *first model* of his telegraph, made in 1837, for the Patent Office.

[39] Daniell's Introduction to Chemical Philosophy, page 580, 2d Edition, London, 1843

www.ingramcontent.com/pod-product-compliance
Ingram Content Group UK Ltd.
Pitfield, Milton Keynes, MK11 3LW, UK
UKHW031350260325
456749UK00003B/516

9 789366 388069